Natural Medicine from Honey Be

(Apitherapy)

Cover: Honey Bee carrying pollen on the hind legs.

Natural Medicine from Honey Bees (Apitherapy)

Jacob Kaal

Kaal's Printing House

First published in Holland under the title 'Apitherapie'
by Kaal's Printing House
© 1987 Jacob Kaal
English edition, updated and revised
© 1991 Jacob Kaal

Dr Bertie Kaal
Singel 368 hs
1016 ah Amsterdam
Netherlands

Published 2017 by:
NORTHERN BEE BOOKS
www.northernbeebooks.co.uk

CIP-gegevens Koninklijke Bibliotheek, Den Haag

Kaal, Jacob

Natural Medicine from Honey Bees: propolis, bee venom, royal jelly, pollen, honey, apilarnil / Jacob Kaal ;
[transl. from the Dutch by Michael Collins].
- Amsterdam: Kaal's Printing House
Vert. van Apitherapie. Amsterdam Drukkerij Kaal, 1987.
- Met index, lit. opg.
ISBN 978-1-912271-08-5

Trefw.: Honingbijen / Natuurgeneesmiddelen.

Preface to the reprinted edition

Jacob Kaal's first booklet, Apitherapy: Curing with bee products, appeared in 1986 (in Dutch). It was updated and translated into English in 1991 and is now reprinted in the original. The book contains an impressive international bibliography and a systematic, detailed record of known and potential medicinal properties of bee products. Propolis had become a forgotten substance in the West for over a century, until the 1960s, and Jacob was determined to promote its varied medicinal powers. Being a beekeeper, his research always touched ground with the life and wellbeing of bees and people.

At an early age Jacob Kaal learned beekeeping from his mother. He soon joined the Dutch Beekeeping Association and became a beekeeping teacher and later chair of the Amsterdam-Amstelland Beekeeping Association. He combined these activities with his passion for printing and nature in general. From 1961-1964 he took the whole family, including five children between 5 and 14 years-of-age, on an adventure to Tanganyika where he taught beekeeping to local Wachaga people, as an officer of the Queen of England. Working closely with the local experts, he started experimenting with African and European bees and learned that apitherapy was real medicine in Africa, rather than 'alternative'. Back in the Netherlands he was involved in the organisation of the Netherlands Expertise Centre for Tropical Apicultural Resources (NECTAR). His observations were unique and are remembered in current developmental beekeeping projects, such as Bee Support.

In the 1980s Jacob Kaal was part of a wide European apitherapy network and launched his Bee Product Centre. He developed recipes for natural bee medicine in his home laboratory, leaving a strong smell of propolis around the house when he was experimenting with drying, grinding and mixing substances. Whether he was traveling or at home, people kept coming and going to deliver new products, to discuss apitherapy, or to find a cure for their ailments. His wife Janneke supported his work by welcoming people and taking care of the business end.

Connections with an Amsterdam hospital led to developing medicine to relieve AIDS patients from infections. As he became more confident he flew out to meet like-minded beekeepers at Apimondia Congresses and apitherapy centres. His contacts reached from Amsterdam to Paris, London, Basle, Bucarest and other places in Eastern Europe. He would return with stories of his knowledge exchange and bulk orders for raw propolis, honey, royal jelly, pollen, bee venom and apilarnil. In 1989 he produced a catalogue of his products, Bee Health Book (in Dutch), which explains the properties of his tablets, capsules, powders, ointments, tinctures, creams and pastes, as well as his famous Ambrosia wine that is still being produced. The booklet includes a table listing common ailments and the appropriate product/quantity to be prescribed after consultation. The business was running so well that his sons and daughter were often found in his lab pressing capsules or grinding propolis. His granddaughter, Martina, grew up with a 'fragrance' of propolis and a taste for honey. She learned many practical things from observing her busy-bee granddad, while experiencing the benefits of a propolis ointment he prepared for her dry skin.

The meticulous descriptions of the properties of bee products and the useful summaries of existing literature have inspired many beekeepers across the world. Even today, when his children and grandchildren travel abroad we meet people in far-off places who ask if we are perhaps related to Jacob Kaal, and where can they find his book? We are therefore delighted that it is now reprinted by Northern Bee Books in its original form, providing a historic document of developments in natural bee medicine. The practical perspective taken and the pioneering research make it a valuable source of information for beekeepers, medical practitioners, pharmacists and patients seeking alternative medicine, even after twenty years. But then again, what are two decades in the light of thousands of years in which natural bee medicine has been used for health purposes? We very much hope that this book will contribute something to taking apitherapy out of the 'alternative' sphere, inspiring both awareness of the importance of bees as well as further research for effective treatment with natural medicine from honey bees. It's the bee's knees.

In memory of Jacob Kaal (2-7-1921 to 1-1-1994)
Bertie Kaal, Amsterdam, 2 July 2017

Contents

Foreword

Man's use of substances produced by bees goes back tens of thousands of years. Only simple tools - and a modicum of courage - are required to obtain food from bees. And not just food. The bees produce foodstuffs such as honey, pollen and bee larvae. And in addition, a substance known as propolis, which has been used as a medicine since the time of the early Egyptians. Propolis and honey contain selective antibacterial substances. They also halt or destroy a whole range of infectious or toxic fungi. And they do this without affecting the healthy intestinal flora.

Bee products go through a highly complex manufacturing process. In order to produce one jar of honey a colony of bees has to travel a distance equal to a threefold circumnavigation of the earth. The transformation of nectar into honey requires that between 30 and 80 % of the water content must be removed by evaporation. Collecting pollen is a herculean task and pollen needs to be sterilized and stored in the combs. Last but not least, propolis has to be collected by the senior workers. Propolis is a sticky substance that bees collect on their hind legs - just as they do with pollen. Sometimes two or three bees are required to scrape the propolis off the legs of the old worker bee.

Bees, plants and man work in very close harmony. We only need to call to mind the many types of fruit that we would not have if there were no bees to pollinate the blossom. Fruit and many types of plants are essential to the health of human beings. In return people provide a safe environment for bee colonies and from early times beekeepers were assiduous in keeping their honeybees alive. Bees assist in the propagation of plants. And, at the same time, they make various products, most of them with a nutritious and/or medical quality for both, humans and animals. These substances produced by bees are honey, wax, pollen, bee bread, propolis, royal jelly and bee venom.

Since the 16th century man-made products have begun to replace such bee products as honey for sweetening and wax for light. This development has made bees less important in our society as producers of useful materials. We now realise that natural products often have more sophisticated qualities.

In the late 19th and early 20th centuries the medicinal properties of bee products had been forgotten. Until 1960 it was impossible to find any mention of propolis in the pharmaceutical manuals. And yet, the Egyptians and Greeks had regarded it very highly as a medicine. Even today, despite all the research that has been done, there are very few doctors who know about the medicinal properties of propolis. The medical world remains aloof from alternative types of therapy, something which is a source of surprise to many natural healers working succesfully with natural products.

The disappearance of propolis from the medical stage is even more of a puzzle when we consider that, for instance, it was used on a large scale during the Boer War in South Africa. There are reports dating from this period which tell of the success obtained with propolis in the treatment of traumatic fever and gangrene. Propolis was used at the time because there was a shortage of 'recognized' medical supplies. Experience showed that it was more effective than the other remedies. Fortunately many universities have now initiated research into substances produced by bees.

I am grateful to the Royal Society of Medicine, London for their permission to print excerpts from an article on antibacterial properties of propolis (page 41), and to Birkhäuser Verlag, Basel to reprint parts of a study on propolis and its cytotoxicity on tumor cells (page 22) and I thank Prof. Dr S. Scheller for his kind permission to print excerpts from articles on propolis and its effects on gamma irradiation and from an article on the antitumoral property of propolis (page 28). He also provided me with up-to-date information about recent research on the anti-oxidant property of propolis and its radical scavenging ability (page 30).

And finally I wish to thank members of my family: my daughter Bertie Kaal MA for her scrupulous work on the manuscript and the proofs of this book and my son Frits Kaal who is responsible for the typography and printing of this book. He made its appearance very much to my taste.

Jacob Kaal, Amsterdam, September 1991

Propolis

Propolis as a curative agent

Propolis is a kind of balsam collected by bees from the buds and leaves of different types of trees and plants. The bees mix this balsam with substances derived from pollen and different types of active enzymes. The enzymes are secreted from glands situated in the head and thorax of the insect. Bees have various uses for propolis: (1) it is added to the wax combs in small quantities; (2) the whole bees' nest is enveloped in it for protection and storage; (3) the cells where the queen is to lay her next batch of eggs are thoroughly polished with propolis so that the larvae are protected against disease from the moment they hatch out.

Apparently propolis is very selective in the way it attacks pathogenic bacteria as well as mycosis and virus infections which can be dangerous to human beings. And because of its selectivity the normal bacteria, such as the flora which perform a vital function in our digestive system, are not affected. Propolis does not cause any harmful side-effects (except in rare cases of allergy) nor do bacteria or viruses build up resistance to it. This means that propolis can be used repeatedly in the treatment of illness, unlike the penicillin type of antibiotics which has limited effectiveness because bacteria gradually develop immunity to it. If penicillin is used, propolis renders it more effective. Likewise the use of propolis usually enhances the effects obtained from other modern drugs.

A cautious approach to the use of propolis is advisable as approximately one in every 2000 people show an allergic reaction to the substance. Regular use can also cause the build-up of an allergic reaction. On the skin this is signalled by the appearance of a pale red patch where the skin is thin or sensitive. The patch disappears within a few days. Any itchiness suffered can be quickly stopped by applying a mixture consisting of four parts glycerine to one part ammonia. When propolis is taken internally an allergic reaction is much rarer, but it can manifest itself by the appearance of tiny red spots on the chest with some slight itchiness. In such cases it is advisable to use propolis in a very diluted form - or to stop the treatment altogether.

Quite likely everyone has their own sensitivity to propolis. And thus each individual can best decide for him/herself which concentration works best. Propolis should not be used in concentrations greater than is required. This booklet gives general indications as regards dosage, but it is quite possible that a half or a quarter of the recommended dose is sufficient. It is a question of testing it out on yourself.

In the hive propolis plays an extremely important role since it protects the honeybee colony from practically every undesirable microbe, fungus and virus: it either kills off or prevents the spread of infectious agents. If a mouse manages to get into the hive it is first stung to death by the bees and then completely encased in propolis - as if embalmed - after which it slowly desiccates without actually decaying and leaves no poisonous substances behind. Even the odour of propolis can kill bacteria. Beekeepers are well aware of how a blocked nose clears almost immediately once the smell of a bee colony has penetrated the nose when the hive is opened for inspection.

Beekeepers have nick-named propolis 'bee glue'. The bees use it when they start building a comb because (as we think) it is tougher than wax and it strengthens the structure. Also a small percentage of propolis is mixed in the wax. The bees use it to close cracks in the wall of their hive, sometimes they make the entrance smaller with the aid of propolis and they often cover completely with propolis the upperscreen, fitted by the beekeeper to provide ventilation, for the bee colony.

The hive requires the presence of an extremely active substance in order to combat sickness since honey bees live tightly packed together not only at night and when the weather is bad but also throughout the winter. In periods of extreme cold the bees huddle together as closely as possible in order to conserve body heat. In such conditions, quite obviously, infectious diseases would soon gain the upper hand if the bees were not well protected.

10 Thanks to its effectiveness in combating bacteria, mycoses and

viruses propolis is important as a protection against infection for the whole bee population.

Propolis provides the same protection against bacteria which cause disease in humans. This fact has probably been known since the dawn of humanity. Propolis was described by the Romans and Egyptians and has always played an important role in traditional folk medicine. And there is clear evidence of a growing interest in propolis on the part of modern medicine. Propolis is a natural substance that has been used for thousands of years as a curative agent; it is now being studied by more and more doctors, bacteriologists, chemists and biologists and it promises to provide still more types of therapy in the future. It is at present not only used for medical purposes, but also in cosmetic creams because it strengthens the tissue and stretches the skin. It can also be used in cleansing emulsions as a therapy for acne and as a face mask. Propolis cream combined with propolis tablets or capsules for oral use is very effective against various types of eczema (psoriasis, for instance).

Propolis is the most effective means of applying plant-derived therapeutic substances to the human body. While honey and pollen are often very effective against illness with non-specific symptoms, propolis can be used specifically against infections - both internal and external- skin conditions, swellings and wounds. In illnesses consequent on diminution of function in the endocrine glands or other organs propolis can work as a stimulus, providing the body with greater resistance because glands and organs are stimulated into higher activity. Disorders (including such things as skin conditions and oral infections) can simply disappear. In the case of oral infection it is recommended that propolis be supplemented both internally and externally, where necessary, with as small a dosage as possible of conventional medicines. And here the patient must be alert to any possible side effects arising from conventional medicines. Propolis is a natural substance with no other side effects than the odd case of allergic reaction in the form of a rash which quickly disappears. There are even cases of propolis working too well in certain individuals, which means that some patients require only a weak dilution (1 part to 10 or even 100). Exclusively internal use helps in the case of patients with an allergic reaction to propolis on the skin.

The principal components of propolis are active essential oils (flavonoids) derived directly from the buds of trees and plants. The resinous raw material is mixed with bee saliva. Fermenting agents are added as are three types of vegetable wax all of different weight. Propolis is a very stable substance, even if it originates from different geographic areas with different flora and climates. It contains substances which are relatively equally present. The relative quantities are subject to variation. Substances available in large quantities include various flavonoid groups, (betulinol, quercetin, isovanillin), caffeic acids, unsaturated aromatic acids and ferulic acids. Waxy components make up approximately one third of crude propolis.

Flavonoids can be found everywhere in the plant world but in propolis they are present in a concentrated form. The bees are instrumental in passing on active vegetable substances to mankind (and to animals). The therapeutic use of plants is thus extended and rendered more effective with the aid of the bees.

How flavonoids work

Flavonoids principally affect the cardiovascular system, physical weakness, permeability and blood flow. They also lower blood pressure and dilate arteries. Other effects noted: diuresis, improved gall excretion, stimulation of various endocrine glands such as the thymus, thyroid, pancreas and adrenal glands.

Ferulic acid

This substance is found particularly in larch, pine and fir tree resins. Bees obtain it in the form of a glycone that is combined with sugars to form a glycoside. Ferulic acid is characterized by a strong antibiotic effect (against gram negative and positive organisms). It also has definite agglutinating (blood-clotting) properties which come in useful in the treatment of wounds that heal with difficulty, being administered in the form of an ointment containing propolis dissolved in alcohol. The ferulic acid component makes propolis effective against collagen disease.

How propolis works

Propolis is antibiotic, antivirus and antiparasitic; it also provides protection against radiation.

It is versatile in its effectiveness against skin conditions, urinary tract infections, prostate problems, disorders of the endocrine system, local pain (including toothache) and in the treatment of wounds. As far as the latter are concerned, propolis provides a local anaesthetic and it heals wounds leaving practically no scar.

Dosages: test results

In his book 'La Propolis' (Paris, 1980) Dr Yves Donadieu writes of the use of propolis as follows: "Propolis was administered at a high rate of dosage (10 to 15 grams daily per kilogram of body weight) in tests on animals - including dogs, rats and guinea pigs. Even after several months of testing, no toxic effects were noted. The high dosages led to no pathological disorders."

Propolis has not proved carcinogenic in animal experiments nor does it cause tumours. In certain experiments tumours have been observed to shrink or the diseased tissue has been encapsulated when propolis was administered. No foetal damage following the use of propolis has been reported in any of the literature. In general human beings react favourably to propolis. There are a few exceptions to this (rare allergic reactions) but there are no other side effects.

Characteristics

We have learned of the innumerable medicinal characteristics of propolis mostly through empirical observation. Indications that have now been determined through the application of the scientific method will be given in detail. Below is a list of the principal characteristics of propolis:

1. The capacity to destroy or halt the multiplication of numerous types of bacteria. This applies to groups such as the staphylococci, the streptococci and the salmonellae; *bacillus subtilis, -alvei* and *-larvae; proteus bulgaris* and *escherichia coli B*. Experiments have shown that propolis is more effective when applied in or on the body then laboratory 13

experiments show (it is more active *in vivo* than *in vitro*), probably because propolis stimulates the functions of a number of different organs and thus induces the body to show even more resistance to the causative agents of disease. The effects included in this category can probably be attributed to such substances as benzoic acid, ferulic acid, galangin and penocembrin.

2. The antimycotic properties of propolis are known. The substances responsible for the effect propolis has on mycoses include caffeic acid, benzylcumarate P, pinocembrin and pinobanksin.

3. Propolis has anaesthetic properties, stronger than cocaine for local anaesthesia and much stronger than novocaine. This effect is produced by the essential oils.

4. Propolis is effective against virus infections such as those of influenza and herpes but also, for example, against the virus of tobacco mosaic disease.

5. Propolis is very effective in preventing and curing inflammation.

6. Propolis stimulates the formation of cells and tissue. This is an important characteristic in promoting the quick healing of wounds and it diminishes the occurrence of scars.

7. The anti-rheumatic properties of propolis are particularly evident when it is used internally.

8. Propolis stimulates the immune system and strengthens resistance to infectious diseases in the body.

9. Propolis counteracts the oxidization of substances subject to decay (fish, wine, meat, etc.) and could thus be of use in the food and cosmetics industries.

10   Propolis is a phyto-inhibitor, preventing the germination of seeds and the growth of plants. Use could be made of this property as an alternative to gamma radiation applied to prevent germination in garlic, onions, shallots and potatoes.

Composition

In order to analyse propolis it needs to be dissolved. About 60 % of crude propolis resin dissolves in pure alcohol (96 %). If a 70 % strength alcohol is used the various waxy components are not dissolved. They form a deposit at the bottom of the container and can be removed. This is better as regards the treatment of wounds since the particles of wax slow down the effectiveness of propolis. A mixture of equal parts of acetone, alcohol, ammonia, benzene, chloroform, ether and trilene dissolves propolis entirely.
Propolis consists of the following:

approx. 50 % resins and balsam
approx. 30 % vegetable waxes
approx. 10 % essential oils
approx. 5 % pollen
approx. 5 % organic substances and minerals.

Laboratory analysis has revealed that propolis contains many components, including: acetatin, caffeic acid, phenylacrylic acid, ferulic acid, phenylacrylic alcohol, chenipheride, chrysin, dimethoxy flavonoid, galangin, luteolin, kaempferol, arigine, proline, isovanillin, isalpinin, benzylcumarate P., pectolinarin, genin, pinobanksin, pinocembrin, pinostrobin, quercetin, rhamnocitrin, sukaranetin, tectochrysin and vanillin.[1]

Among these substances the flavonoids in particular have considerable effect on the growth of capillaries, they reinforce the effectiveness of ascorbic acid and they work against all types of infection.

Propolis is rich in minerals, but because the amount of propolis required as a daily dose is small, the mineral intake is not sufficient

1   Data on the methods used to investigate the presence of these substances can be found in 'Propolis' (Apimondia; Bucharest, 1976); 'Plant Origins of Propolis: A Report of Work at Oxford' (Bee World 1990); Scheller et al., 1990, 1991, (see page 30 of this book) and König, 1986, see pages 21-22 of this book.

for the body's daily requirements. However minerals taken via propolis do make a positive contribution. (Some minerals are only required in trace amounts.) The following minerals have been established as components of propolis: aluminum, barium, potassium, chromium, cobalt, copper, tin, iron, manganese, nickel, lead, silica, titanium, vanadium and zinc.

Using propolis
Propolis can be administered orally as follows:

A.   Healthy people can take oral doses of propolis in order to increase their natural resistance. However, we should not take too much - a maximum of about 30 mg per day, since there is no illness to combat. Elder people can take as much as 60 mgr daily to avoid 'flu' and other infections.

B.   Sick people can often take propolis without any additional preparation, though it works wonderfully well in combination with other medicines or with plant extracts and essential oils. In almost all cases the effect of other medicines is enhanced when taken in combination with propolis.

Therapeutic use

I.   The cardiovascular system and blood circulation

II.   Ear, nose, throat and bronchial tract. This is a preferred terrain for propolis and it proves effective in both acute and chronic cases. Used principally for:
throat infections;
pharyngitis, rhinopharyngitis;
laryngitis;
rhinitis (particularly ozena);
sinusitis;
otitis.

III   Bronchial and pulmonary ailments in general. Propolis is particularly effective in the treatment of tuberculosis.

IV   The digestive system
stomatology (oral infections);

dental hygiene, halitosis;
infections of the gums and tongue, ulcers, aphthae;
toothache and inflammation;
moniliasis (candidiasis - Candida albicans infection)
gastritis, gastric and duodenal ulcers;
colitis;
Crohn's disease;
infectuous - mononucleosis (Pfeiffer's disease)

V   Genitals and urinary tract
bladder and kidney infections;
infections of the genitals (particularly male prostate problems and female vaginal trichomoniasis).

VI   Dermatology (usually treated locally, sometimes supported by oral intake of propolis tablets or capsules):
bruises, cuts;
frostbite, chilblains;
first and second degree burns (including sunburn);
abscesses, suppurating wounds, boils, varicose veins, ulcers stemming from bad circulation;
in general all slow-healing wounds and awkward scarring;
diseases of the rectum;
corns, callouses on the sole of the foot or palm of the hand, horny skin, warts, infected scar tissue, skin discolouration (in infants: intertrigo), eczema, psoriasis;
radiation-induced skin infections;
mycoses (including ringworm); tinia of the feet (athlete's foot)
shingles (good chance of healing with propolis).

VII   Stimulation of the endocrine system (especially in cases of goitre).

VIII Various rheumatic syndromes affecting the joints.

IX   Ophthalmology, various types of inflammation (styes, blepharitis, etc.).

17

Effective treatment of the conditions listed above is based mainly on practical experience. A great deal of research still needs to be done before scientific proof is available. And yet, propolis has clearly demonstrated its merit since its use has led to satisfactory results, often more quickly and more effectively than conventional medicine. Propolis often works where other remedies have failed. The medicine can be regarded as an agent in the healing process.

Dosage and applications
1. Crude propolis can be used in its natural form (propolis crumbs or pieces). It can be warmed and softened in the mouth and then pressed into place on, for instance, a tooth. It is then slowly dissolved by saliva and absorbed into the body. Propolis used in this way, for inflammation or ulcers in or on the mouth, throat, uvula or larynx, is very effective since it penetrates the mucous membranes immediately. It also relieves toothache because of its local anaesthetic effect. Crude propolis can also be taken in capsule form. It will be dissolved in the stomach and becomes active throughout the body.
2. Extracts of propolis diluted in alcohol in 3 % - 30 % concentrations can be used directly on wounds, burns and sometimes for persistent eczema. Such extracts can also be taken orally, e.g. a few drops on a spoonful of honey.
3. Creams and salves, consisting of propolis mixed with a greasy emulsion (such as lanolin) or in the form of a jelly (non-greasy). Here the propolis concentration can vary from 1 % - 20 %.
4. Purified propolis, with the waxes and contaminants separated from the resins and essential oils by cold centrifuging. It can be dissolved in alcohol with a concentration up to 50 %. Purified propolis can be used in salves, creams, toothpaste, tablets, etc.

Many propolis products are manufactured and marketed all over the world. The recommended doses vary widely. In the case of crude propolis, for instance, which still contains 30 % of vegetable waxes, it is recommended that between 1 and 3 grams per day be taken for a period of 20 to 30 days. This means that 2 grams of purified propolis could be required as a daily dose. For the tinctures, containing 20 % propolis, the prescribed dose is 15 drops per day. If propolis capsules containing crude pulverized propolis is preferred, a daily dose of 2 to 6 capsules (0.20 gram per capsule) can be maintained for 3 to 4 weeks.

In severe cases like for example 'erycipelas' it is possible to take 12 capsules per day for about a week.

For chronic illnesses - rheumatic pain, gall and liver conditions, influenza, hoarseness, inflammation in the head - tablets containing 60 mg of purified propolis can be taken (about 3 - 6 per day): they are dissolved in the mouth and slowly swallowed with the saliva

The following preparations are available:

1. **Propolis tablets** in blisters containing 40 tablets per box of 60 mg extract of 40 % purified propolis in 70 % alcohol.

2. **Propolis ointment** in three varieties: mild (approx. 7 % purified propolis), medium (approx. 12 % purified propolis) and strong (approx. 15 % purified propolis).

3. **Propolis capsules** in plastic jars containing 30 capsules of 200 mg (0.2 grams) of cleaned crude pulverized propolis. Propolis cannot be pulverized at room temperature so it is first frozen and then ground to a fine powder which is mixed with a few per cent of protein material - such as soy meal or pollen - to make the propolis more acceptable to the stomach.

4. **Propolis spray** (10 % purified propolis extract) in 15 cc cans mainly for use on 1st and 2nd degree burns and also on healing wounds to prevent the formation of scar tissue. Any wound thus treated does not need to be bandaged since the propolis dries out and forms an elastic, porous membrane on the skin. Propolis spray is also extremely effective in regenerating and rebuilding tissue cells and this causes burns to heal rapidly. The spray not only disinfects but also acts as an analgesic. It is equally recommended for skin diseases such as eczema, proud flesh, ulcers, escherichia trychophytes, staphylococci. The spray can be used very effectively for emergency treatment.

The area to be treated is sprayed. The force of the spray pushes the preparation into the pores and provides the damaged skin with a thin protective membrane which prevents further infection. The regeneration of tissue and the closing of the wound are accomplished under the protective layer of propolis.

5. **Propolis tincture in distilled water**. This tincture is made from propolis by dissolving the propolis in alcohol so that thefrom which resins and vegetable waxes are to be removed. The essential oils which remain are the principal ingredients. The tincture is then mixed with distilled water, leaving a 40 % (approx.) alcohol solution. It is a clear liquid which immediately turns milky on being mixed with water: this is caused by certain microscopic paticles invisible in the alcohol solution, which but become visible when the proportion of alcohol is reduced. A strong solution such as this is eminently suitable for dilution with water.

Application 1: Propolis in this form can further be used as an antibacterial, antimycotic and antiviral rinse against diseases of the mouth (stomatitis, aphthous ulcers). For use as a mouth rinse, it is sufficient to mix a few drops with water in a small glassand then rinse the mouth. The strength of the solution is determined by the user on the basis of experience.

Application 2: To combat viruses, bacilli and fungi which may establish themselves on and around the eye: a few drops (3 - 5) in an eye bath filled with tepid boiled water provides sufficient strength.

Application 3: The liquid is also effective for intestinal and stomach disorders when taken orally, 10 drops in a small cup of water, to be drunk in the course of a day.

Application 4: A solution of five parts water and one part tincture i effective when used on the skin to stop itching or for any dermatitis. For this purpose it can be used twice a day.

6. **Propolis tincture** consists of a 20 % purified propolis solution in 70 % alcohol. It is used for grazes, burns, dermatitis, festering wounds, infections, etc. It forms an elastic membrane over the wound, making a dressing superfluous. Tissue growth is encouraged (practically no scars are left), bleeding is stopped and pain relieved.

Propolis tincture can also be taken orally: a few drops, for instance, on a teaspoonful of honey.

7. **Propolmel** is a very valuable product consisting of 70 % cold extracted honey, 20 % pollen and 10 % propolis. Propolmel is mainly used for stimulating glands, against infections and haemorrhages of the stomach and intestine and for the treatment and prevention of prostate conditions. Pollen tends to raise blood pressure but honey restores the balance. A new preparation being developed is known as **Propolmel plus**, to which royal jelly, lotus and ginseng have been added. When prescribed for the elderly, for the chronically sick, for post operative convalescents, for patients requiring a post-operative tonic, or for those suffering from anaemia and undernourishment this preparation constitutes a physical and mental tonic. It is generally very effective in building up resistance, also against gamma-irradiation.
Other preparations available on the market are:

8. **Propolis cream**, containing 1 % or 2 % propolis and used for skin treatment. The contractile strength of propolis combats the formation of wrinkles on the face and causes liver spots to disappear. It can be used to combat fungal diseases of the skin and is also prescribed successfully for psoriasis, in combination with oral intake of propolis. In most cases the cream seems to work well when the patient cannot take the stronger salve (causes itching). If long-term use of propolis on certain areas of the skin has led to a propolis allergy, it is still often possible to use the propolis cream.

The following is a selection of abstracts from scientific research published over the past ten years. They will give an impression of current research and attitudes towards propolis in different parts of the world as well as the large variety of applications.

König, B., (1986) **Studien zur antivirotischen Aktivität von Propolis (Kittharz der Honigbiene, *Apis mellifera*)**, BSc Dissertation, Hannover.(In German)

Summary
Propolis or bee glue of different geographic origin has been extracted according to various described methods, the percentage of total weight of the individual fractions (ethanol-soluble, balsamic, lipophilic or wax and the insoluble remainder) has been determined as well as the physiocochemical data of the ethanolic solutions and the citrate 21

containing aqueous suspensions (pH, refractory index, electric conductivity). Poplar bud extractions were treated in the same way.

Among the analytical methods applied (TLC, GC, GC-MS, HPLC), HPLC proved to be the most suitable. The results obtained by the other methods previously mentioned proved to be of a more limited value. Their benefits as well as their drawbacks are discussed in detail.

Caffeic acid and other cinnamic acid derivatives are shown by those methods to occur in all non-tropical samples. Among the flavonoids 7-methoxyquercetin was found in nearly all of these samples, in the majority of cases accompanied by 3,7-dimethoxyquercetin, and traces of quercetin and luteolin were also generally present. With HPLC it was possible for the first time to observe dicaffeoylic compounds in propolis (detected by UV-spectra combined with high retention periods). However their structural elucidation is still pending.

With several herpes viruses known to be pathogenic to birds and mammals, clear cut inhibition effects of all tested propolis-samples from non-tropical regions were observed and quantified. However in all samples except one (sample XIX from Ithaca, Greece), cytotoxic activity was observed in the concentration range immediately above the antivirally active concentrations. The Hawaiian samples also showed some antiviral activity, but with a narrower range between minimal active and lowest cytotoxic concentration, whereby the cytotoxic activity becomes apparent after a significant longer incubation period than with the non-Hawaiian samples. The reason for the cytotoxicity could not be determined, but it seems to be related to the total content of polyphenols at least in the north-american and european samples.

With the biphasic method developed by the author not only complete propolis samples but also pure known compounds were tested against the viruses. Among them p-cumaric acid, ferulic acid, galangin and apigenin proved very weak antiviral activity and most other compounds, such as e.g. chrysine were inactive.

However, a high antiviral activity is shown by several compounds of different structural classes, which have the caffeoylic molecular moiety of caffeic acid in common. This also proved to be true of some compounds not known from propolis, which were included in my tests, so that I was induced to propose an extension of the term of caffeoylic compounds and to see them as a promising new path in antiviral chemotherapy. This holds true especially when considering that these pure compounds seem to be devoid of any kind of cytotoxicity.

In non-tropical propolis they are represented according to my studies by caffeic acid itself (which previously has been known to be antivirally active, but not to be part of a much larger family of antivirals), by the quercetin derivatives mentioned above, by the newly detected dicaffeoylics and by the traces of quercetin and luteolin, which, together, form the antiviral principal of propolis.

By time-variation experiments it was found, that this antiviral activity acts during virus eclipse and that the mechanism of inhibition may therefore be related to either virus-specific nucleic acid or protein synthesis.

The significance of the results and the possible future guidelines of research in this field are discussed in detail.

D. Grunberger, R. Banerjee, K. Eisinger, E.M. Oltz, L. Efros, M. Caldwell, V. Esteevez and
K. Nakanishi; (1988) **Preferential Cytoxicity on Tumor cells by Caffeic Acid Phenethyl Ester Isolated from Propolis**; In *Experientia* 44. This extract is printed

with kind permission of Birkhäuser Verlag AG, Basel, Switzerland. (Comprehensive Cancer Center, Institute of Cancer Research, College of Physicians and Surgeons, Columbia University, New York and Department of Biological Sciences and Department of Chemistry, Columbia University, New York.)

Summary: The honeybee hive product, propolis, is a folk medicine employed for various ailments. Many important pharmaceutical properties have been ascribed to propolis, including anti-inflammatory, antiviral, immunostimulatory and carcinostatic activities. Propolis extracts have provided an active component identified as caffeic acid phenethyl ester (CAPE), which was readily prepared in one step. Differential cytotoxicity has been observed in normal rat/human versus transformed rat/human melanoma and breast carcinoma cell lines in the presence of CAPE.

cytostatic

The effect of CAPE on human cancer cells was tested by measuring incorporation of [^{3}H]thymidine into the DNA of human breast carcinoma (MCF-7) and melanoma (SK-MEL-28 and SK-MEL-170) cell lines in culture. Figure 1A reveals that 5γg/ml CAPE inhibits incorporation of [^{3}H]T into the DNA of human breast carcinoma MCF-7 by 50 % and is completely blocked by a concentration of 10 γg/ml. More dramatic effects were observed with the two melanoma lines SK-MEL-28 and SK-MEL-170. Figure 1B illustrates the effect of different concentrations of CAPE on the incorporation of [^{3}H]T into SK-MEL-28 cells. At 5γg CAPE/ml the cells displayed minimal incorporation and were completely inhibited at 10γg/ml. Similar inhibitions were observed for HT29 colon and renal carcinomal lines (not shown). On the other hand the effect of CAPE on normal 1434 fibroblast and melanocytes was significant less. Incorporation of [^{3}H]T into these normal cells was inhibited by only 50 % at concentrations of CAPE up to ten times greater (50 γg/ml, data not shown). These differential effects were reminiscent of those observed with the normal CREF versus transformed wt3A cells or with rat 6 versus T24 transformed rat 6 cells.

We have described the identification of a compound present in propolis which is at least partially responsible for its reported cytostatic properties. It represents the most active component as judged by the assay employed in this study (cytostatic towards Ltky cells). Its identification as CAPE, a compound of structural simplicity, permitted a one-step synthesis which was amenable to large-scale preparation. This in turn, allowed a more thorough investigation of CAPE's cytostatic properties in which several differential effects were uncovered. Most interestingly, human tumor cell lines displayed a significantly greater sensitivity to the action of CAPE than analogous normal lines.

Phenetyl alcohol and caffeic acid, the most obvious metabolic products of CAPE, displayed none of the activities mentioned before. Caffeic acid is known to possess several interesting pharmacological properties and one possibility is that it represents the ultimate effector. Esterification of caffeic acid with a lipophilic alcohol may simply facilitate its transport into cells, where it is hydrolized. Ester analogues of CAPE may be readily prepared for testing this and other possibilities.

The ready accessibility of analogues and labeled versions of CAPE will simplify further investigations into its mode of action, and may lead to an understanding of the observed differential effects on a molecular structural level.

Furthermore, such studies with CAPE and other cytostatic compounds may provide a clearer insight into the molecular events responsible for the dissimilar biological properties exhibited by transformed and normal cells. Because the cytostatic action of 23

CAPE is more dramatic on transformed cells, one may reasonably assume that it is at least partly responsible for the claimed carcinostatic properties of propolis.

a)

b)

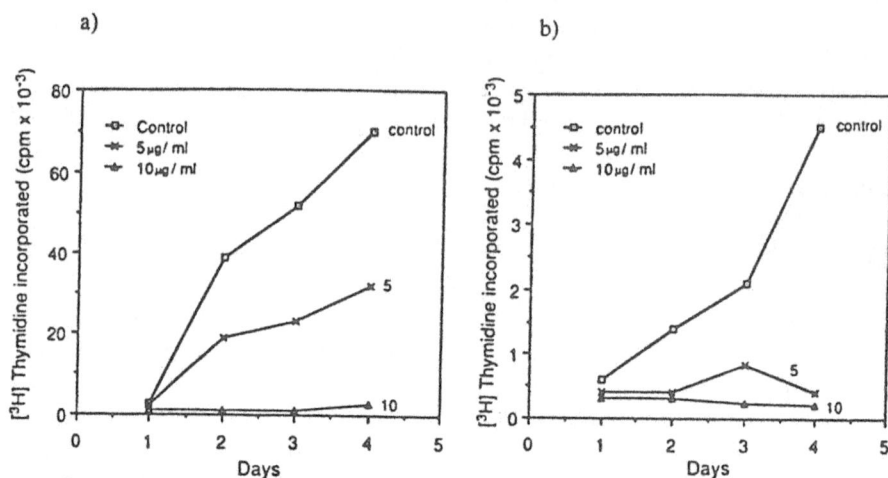

Figure 1. Effect of CAPE on the incorporation of [^{3} H]thymidine into DNA of (a) human MCF-7 breast carcinoma and (b) human SK-MEL-28 melanoma cells. Cells were maintained in Eagle's minimal essential medium (MEM) with Earle's salts and 10 % fetal bovine serum. The cells were seeded in the same medium in tissue culture cluster plates (96 flat bottom wells) at 10^3 cells/well. After 24 h (day 1) cultures were washed and different concentrations of CAPE were added to each well in triplicate. Labeling of cells was accomplished on days 1 - 4 by incubating with 0,5 mCi [^{3} H]thymidine for 5 h.

Stanislow Scheller*, Grazyna Gazda*, Wojciech Krol*, Zenon Czuba*, Alexander Zajusz**, Janus Gabrys*** and Jashovam Shani***** (1989) **The Ability of Ethanolic Extract of Propolis (EEP) to Protect Mice against Gamma Irradiation.** In Zeitschrift für Naturforschung, 44c, 1049-1052. This extract is reproduced with kind permission of the Head of the Institute of Microbiology Silesian School of Medicine: Prof. Dr S. Scheller. (Department of Microbiology* and Department of Histology and Embryology***, Silesian School of Medicine, Zabrze-Rokitnica, Poland; Institute of Oncology**, Gliwice, Poland, and School of Pharmacy*****, University of Southern California, Los Angeles, California 90033, U.S.A.)

Ethanolic extract of propolis (EEP) was tested as a protective agent against gamma irradiation in mice. The mice were exposed to 6 Gy gamma irradiation from a ^{60}Co source, and were treated intraperitoneally with EEP, administered before and after their irradiation. While the non-treated mice expired within 12 weeks, the mice that received a series of EEP treatments survived the irradiation, and their leucocyte count as well as

24

their spleen plaque-forming activity returned to normal. It is suggested that an antioxidant and a free radical scavenger in the EEP are responsible for the radiation protective effect of the extract of this natural product.

Introduction

Propolis is a resinous substance produced by honey-bees. Previous publications from this laboratory reported some features of propolis which are of medical interest and of its ethanol extract (EEP) and demonstrated their immunological properties in laboratory animals and in patients. In experiments carried out in mice it was shown that if the animals' immune system had been triggered by sheep red blood cells, EEP intensified the immunization process, as evaluated by increase in the number of their plaque-forming spleen cells. In the clinic we have shown that in aged people with impaired immune system, EEP revitalized their immune system remarkably.

Recently we noticed a rise in the lytic capacity of some human cell lines preincubated with EEP (unpublished). We also noted higher survival of mice with Ehrlich ascites carcinoma, pretreated with EEP. Some of these properties of EEP were related to its anti-oxidant and scavenging abilities of free radicals. As oxidation products and radiation-initiated free radicals are major determinants in the viability of some immune mechanisms, people became alert to the appearance of radiolysis products in their bodies. Such products result in DNA fragmentation, chromosomal aberration, hyaluronic acid degradation, lipid peroxidation and diminished contents of proteins containing sulphydryl groups, and may alter the immune mechanism. These changes can be blocked by antioxidants and free radical scavengers. We wondered, therefore, whether EEP could prevent the effects of gamma irradiation on the immune system when administered shortly before or after irradiation of mice.

Results and discussion

The protective effect of EEP against a single 6 Gy dose of whole-body gamma irradiation is demonstrated in the survival curves of the mice (Fig. 1). While none of the irradiated, non-EEP-treated mice, survived the 12th week after their irradiation (group "C") - half of the mice receiving three EEP injections (group "B") reached the 17th week, and all but one mouse in group "A", who were subjected to 15 EEP injections, survived the complete study period.

The regenerative effect of EEP on antibody formation of the plaque-forming cells is summarized in Fig. 2. The number of these cells was reduced to about 20 % of the healthy population in all three groups, but while the survival of the non-EEP-treated mice continued to decline gradually until expiration, this number significantly increased in both EEP-treated groups: group "B" demonstrated a slow survival, but group "A" demonstrated a full recovery, and even significant hyperactivity was noticed (Fig. 2). A parallel situation was demonstrated in leucocyte groups to 10 - 20 % of its initial levels with a continued drop in the untreated group ("C"), a very shallow increase in group "B" and full recovery in group "A" (fig. 3).

From these results it emerges that EEP exerts a distinct radiation-protective effect on mice when administered immediately before and shortly after their exposure to sublethal doses of gamma irradiation. The survival of the EEP treated groups was complete, while the mortality of the non-treated group was 50 %. The formation of 19 S antibodies, as evaluated by the spleens' ability to form plaques, and the leucocyte counts in the treated groups were al significantly higher than in the non-treated controls (group "C").

25

As the active antioxidant in the EEP has not yet been isolated and its rate of elimination from the body has not yet been established - one can only conclude that its t1/2 of elimination in the Tween-80 solvent is longer than 24 h, as apparently some of it was present in the body and exerted the protective effect when the mice were irradiated. The better survival of group "A" as compared to group "B" is partially attributed to the apparent higher level of EEP-derived oxidant in their circulation during irradiation.

The ability of EEP to stimulate production of plaque-forming cells was reported by us recently. The plaque-formation capacity of the spleen cells tripled under EEP treatments from a basic level of 300-400/10^6 spleen cells. The complete recovery of the plaque-forming ability and the leucocyte count of the irradiated mice, as demonstrated in this study, indicate that there are some components in the EEP which demonstrate cell-regeneration properties: the recovery in the plaque-formation ability from 40/10^6 after irradiation to about 440/10^6 are remarkable and highly significant. These findings are in line with our earlier reports that EEP restores the immune system in T-lymphocytes and granulocytes, probably due to its high content of flavonoids. The ability of EEP to activate enzymes *in vivo* and *in vitro* have also to be taken in consideration as a contributing mechanism. The complex mechanism of the radiation-protective effect of EEP is currently being studied in our laboratory.

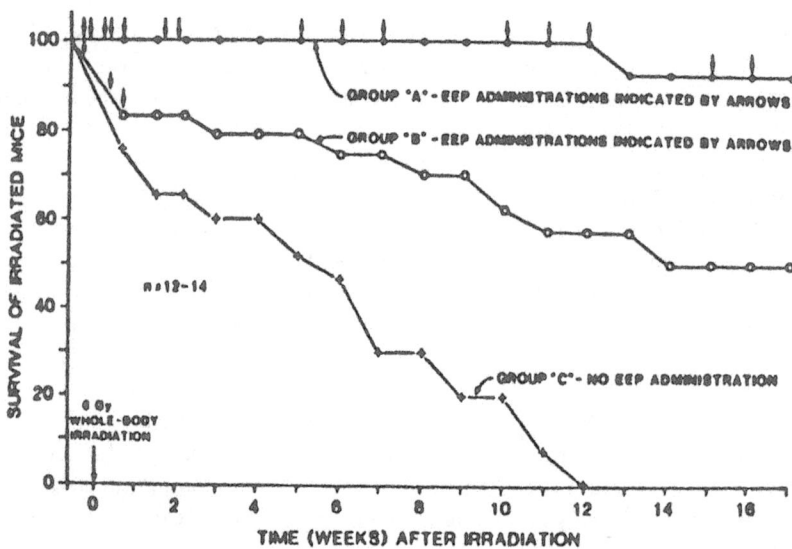

Fig. 1. Survival of BALB/c mice pre-exposed to 6 Gy whole-body gamma irradiation ("0" time), after being treated with 15 (group "A") or 3 (group "B") doses of EEP, as indicated by arrows (n = 12-14).

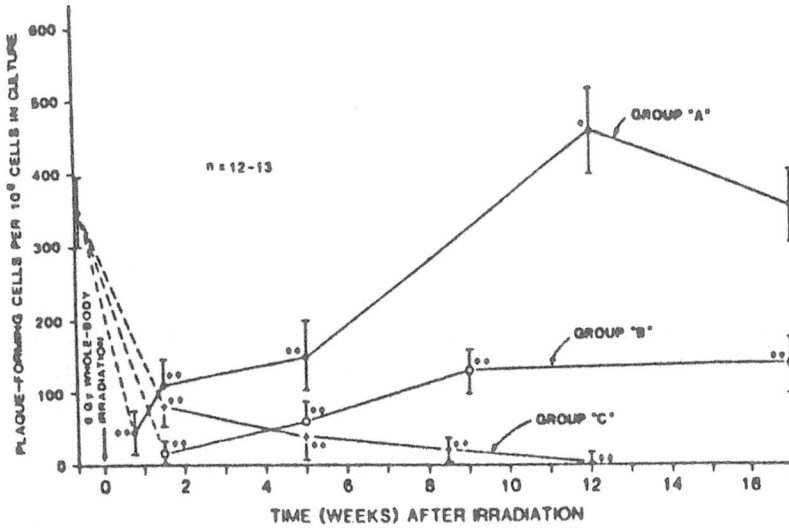

Fig. 2. Plaque formation in the spleens of mice pre-exposed to 6 Gy gamma irradiation ("0" time) and treated with EEP (n = 12-13).

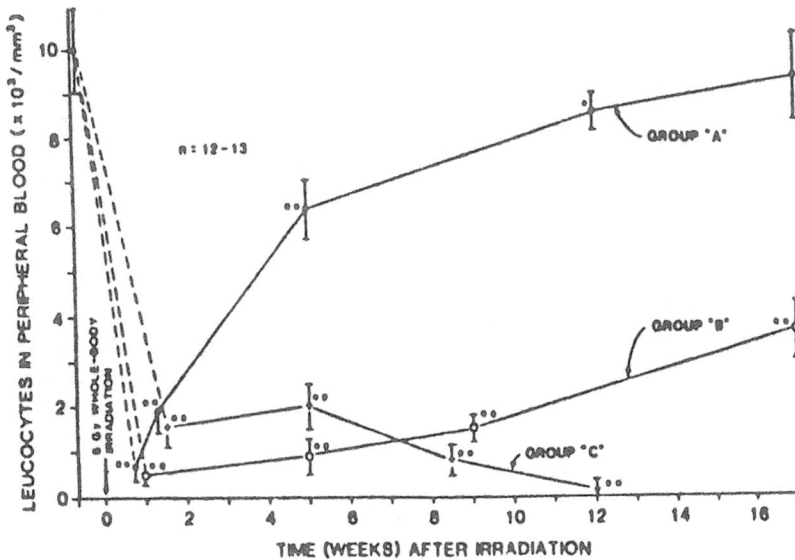

Fig. 3. Leucocyte count in peripheral blood of mice pre-exposed to 6 Gy gamma irradiation ("0" time) and treated with EEP (n = 12-13. *p 0.05; **p 0.01.

S. Scheller*, W. Krol*, J. Swiacik*, S. Owczarek*, J. Gabrys**, and J. Shani** (1989); Antitumoral Property of Ethanolic Extract of Propolis in Mice-Bearing Ehrlich

27

Carcinoma, as Compared to Bleomycin; (Department of Histology and Embryology*, Silesian School of Medicine, 41-808 Zabrze-Rokitnica, Poland and School of Pharmacy**, University of Southern California, Los Angeles, CA 90033, U.S.A.). In 'Zeitschrift für Naturforschung', **44c**, 1063-1065

Antitumoral effect of ethanolic extract of propolis (EEP) was demonstrated in mature mice-bearing Ehrlich carcinoma. Survival rate after EEP treatment was compared to that of bleomycin, given alone or in combination every two days for 36 days and followed up for 14 additional days. The survival rate at 50 days was 55 % after EEP and 40 % after bleomycin, while all the mice treated with EEP + bleomycin combination demonstrated shorter survival than the controls. It is concluded that while the *in vivo* activity of bleomycin is reduced in the presence of cytochrome-C-reductase inhibitors (like some of the EEP components are), the antitumoral property of EEP in the tumored animal model studied is significant and lasting.

Introduction

Propolis is a natural resinous product of honey bees. It is rich in free amino acids and flavonoids, and has antibacterial properties. Its ethanolic extract exhibited marked antiprotozoan activity, it raised the mitotic index of cells cultivated *in vitro*, and intensified NADH2-reductase activity in such cells. *In vivo* EEP demonstrated enhanced activity of the enzymes NADH2 and glucose-6-phosphatase in rats, accelerated the rate of ossification and stimulated regeneration of dental pulp. While parenteral administration of EEP to rabbits does not induce anti-EEP antibody synthesis *in vivo*, it increases the number of cells synthesizing antibodies in vitro. In aging subjects with impairment of immunological functions, application of crude propolis or EEP restored several of these functions. Its immunostimulatory activity was demonstrated *in vitro*: EEP increased the cytotoxicity of NK cells (S. Scheller and W. Krol, unpublished), inhibited the development of HeLa (cervix) and KB (nasopharynx) carcinoma cells. *In vivo* studies indicated that EEP stimulated the immune system in patients with prostate inflammation. These properties led us to compare its antitumorogenic properties in mice-bearing Ehrlich carcinoma with bleomycin, which is highly effective in this tumor model.

Results and discussion

Fig. 1 illustrates the survival of each of the experimental groups, as recorded daily from day 0, through the end of the treatment period (day 36) up to day 50. Survival patterns after EEP and bleomycin were similar, with the effect of bleomycin starting to diminish on day 17, while EEP managed to negate mortality of the tumored mice up to day 25. Survival of the mice undergoing separate EEP or bleomycin treatments were 55 % and 40 % on day 50, respectively. Mortality rates of control mice reached 100 % on day 40-42, while mortality of the EEP + bleomycin combination group was higher than the controls, and reached 100 % on day 33. There is no significant difference in mortality between male and female mice in any of the groups.

28

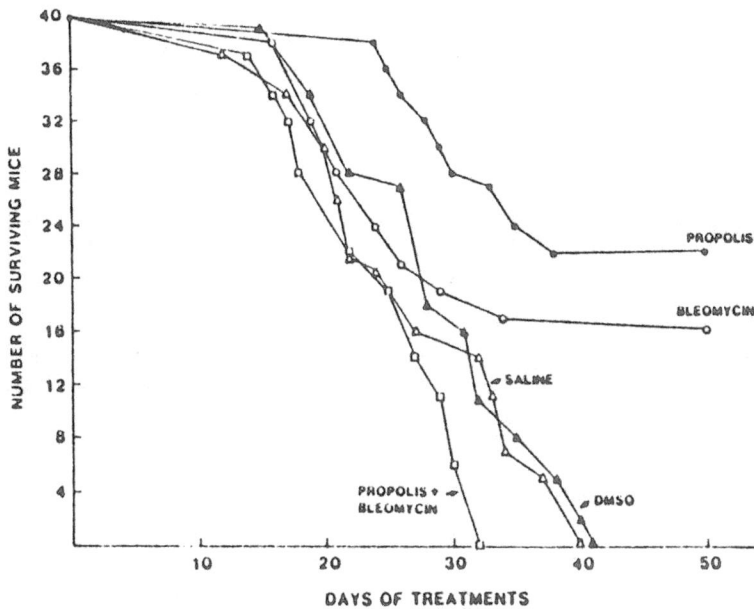

Fig. 1. The effect of ethanolic extract of propolis (EEP), as compared to bleomycin, alone and in combination, on survival rate of mice-bearing Ehrlich carcinoma (n = 40).

Antitumoral activity of EEP against Ehrlich carcinoma cells and the inhibitory effect of different extracts on development of HeLa and KB cells *in vitro* is suggested to be related to its content of flavonoids. Flavonoids affect metabolic stages of Ehrlich carcinoma cells, e.g. inhibit the incorporation of thymidine, uridine and leucine into them which in turn lead to inhibition of DNA-synthesis. The inhibitory role played by flavonoids in the antineoplastic proces, has also been confirmed in other experimental modes. Flavonoids inhibit carcinogenesis induced by polycyclic aromatic hydrocarbons in cancer models. The mechanisms of these activities are connected with the ability of the flavonoids to inhibit metabolic stimulation induced by such polycyclic aromatic hydrocarbons and by affecting the activity of some cell promoters.

Our previous *in vivo* observations suggest that the antitumoral activity of propolis extract is related to immunostimulatory property. Example of this activity is the enhanced antitumoral activity demonstrated by EEP on NK cells and Ehrlich carcinoma cells *in vitro*. The antagonistic effect of EEP + bleomycin combination is probably neutralizing the antioxidative property of the drug. This property is antagonistic to the mode of action of bleomycin, as the latter is a glycopeptide antibiotic, which is activated by superoxide ion and by free radicals, and which *in vitro* exhibits reduced DNA degenerative activity in the presence of standard free radical scavengers. Attempts to elucidate detailed mechanism of EEP activity on Ehrlich carcinoma and further studies on combined therapies using EEP and other antitumoral drugs, are currently being studied in our laboratory.

29

S. Scheller*, T. Willczoks**, S. Imielski**, W. Krol*, J. Gabrys***, and J. Shani**** (1990);
Free Radical Scavenging by Ethanol Extract of Propolis,; (Department of Microbi-
ology*, Biochemistry and Biophysics**, Histology and Embryology***, Silesian School of
Medicine, Zabrze-Rokitnica 41-808, Poland; and School of Pharmacy****, University of
South California, Los Angeles, CA 90039, USA). In Int. Journal of Radiation Biology, vol.
57, no 3, 461-465

The free radical scavenging ability of an ethanolic extract of propolis (EEP) was
demonstrated by electron spin resonance spectroscopy, when 2,2-diphenyl-1-picrylhy-
drazyl (DPPH) was treated with increasing concentrations of EEP. It was shown that the
DPPH signal intensity was inversely related to the EEP concentration and to the reaction
time. It is assumed that the ability of components in EEP to donate a hydrogen atom is
responsible for the lowering of the DPPH-EEP signal, and reflect the anti-oxidative nature
of EEP.

Introduction

Propolis is a natural resinous mixture produced by honey bees. Ethanol extract of
propolis (EEP) has been shown to possess immunological properties in animals and in
patients (Scheller at al. 1988, Frankiewicz and Scheller, 1984). Recently, we noticed
higher survival of mice with Ehrlich ascites carcinoma when pretreated with EEP
(Scheller et al. 1989a), and demonstrated its ability to protect mice against γ-irradiation
(Scheller at al. 1989b). We related some of these properties to the anti-oxidative effect of
EEP. (Krol et al., 1986), and attributed them to the high content of flavonoids in EEP, which
comprises 25-30 % of its dry weight. The flavonoids are relatively well defined qualitatively and quantitatively, and it has been suggested that the therapeutic activity of
propolis and its extracts depends mainly on the presence of flavonoids and their levels.
(Vanhaelen and Vanhaelen-Fastre 1979, Havsteen 1983). Anti-oxidative capacity of
flavonoids was demonstrated also in microsomes and mitochondria of rat liver cells
(Cavallini et al. 1978), in human erythrocytes (Sorata et al. 1984) and in illuminated
chloroplast under aerobic conditions (Takahama et al. 1982). The anti-oxidative effect of
flavonoids has been attributed mainly to their scavenging ability with lipid peroxyl
radicals (Takahama et al. 1984, Erben-Rus et al. 1987).

We suggest that the ability of compounds present in EEP to donate a hydrogen atom is
responsible for lowering the ESR signal of the DPPH, and thereby reflects the anti-oxidant
effect of EEP. Apparently flavonoids are the most abundant and most effective free radical
scavengers in the EEP extract (Krol et al. 1986). The following flavonoids have been
detected in EEP to date: galangin, isalpinin, kaempferol, rhamnocitrin, rhamnetin,
isorhamnetin, quercitin, pinocembrin, pinostrobin, and pinobanksin. In addition, vitamin E, histidine and some additional redox active components of the EEP may take part
in this scavenging proces. Some biochemical aspects of flavonoids and their evolution
are discussed by Swain (1986).

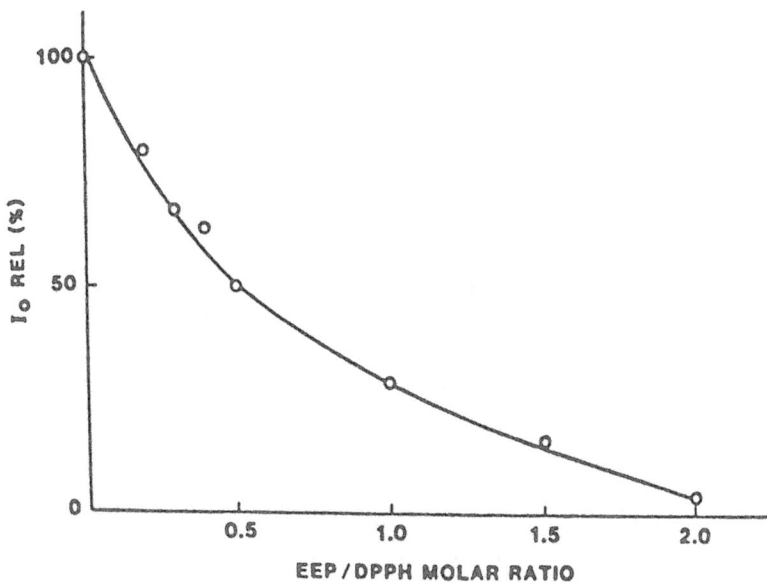

Figure 1. Dependence of the relative intensity of the first component of the ESR spectrum of DPPH, at steady state (4 min after mixing), on the concentration ratio of DPPH and EEP (mean of five experiments).

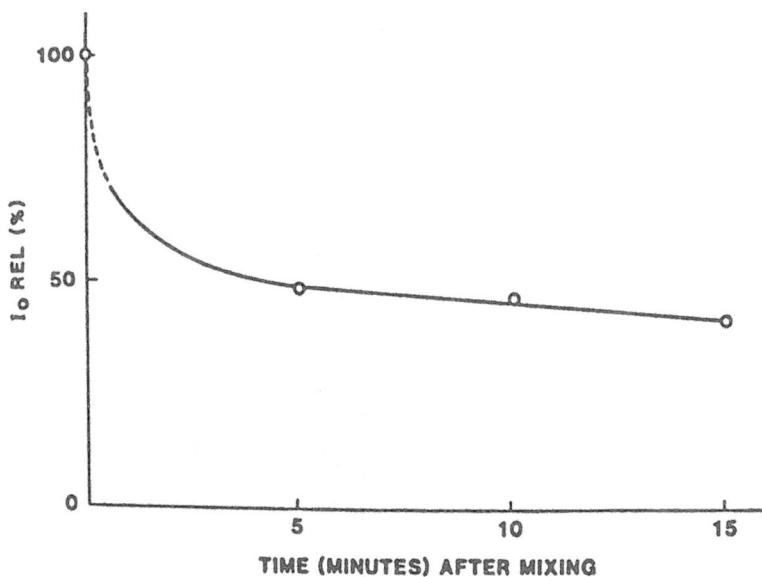

Figure 2. Effect of 0.25 mg/ml EEP on the I_0(REL) of DPPH as function of time (mean of seven experiments).

EEP/DPPH†	I_{INT}	$I_0(REL)(\%)$
0.00	6.8	100
0.14	5.3	78
0.27	4.5	66
0.35	4.2	63
0.50	3.4	50
0.66	2.8	41
1.00	1.3	28
1.50	1.1	16
2.00	0.6	3

† The concentration of DPPH was always 0.5 mg/ml. Ratios were calculated on a weight basis.

Table 1. Dependence of I_{INT} and $I_0(REL)$ of the EEP-DPPH mixture on the concentration of added EEP.

In conclusion, scavenging of a stable free radical, demonstrated here using propolis preparations, might explain some of the therapeutic effects of EEP observed in animal and clinical studies. The most important of these are activation of the immunological system, especially in aged patients (Frankiewiez and Scheller 1984, Scheller et al. 1984), regeneration of bone tissue (Przybylski and Scheller, 1985), regeneration of cartilaginous material, slowing down of tissue detorioration (Scheller et al., 1977) and inhibition of tumor growth in animals (Scheller et al., 1989a). Free radical scavenging, and consequently the anti-oxidative efficacy of EEP, plays a distinct role in some of these processes. The EEP/DPPH ratio of 2-0, where the disappearance of DPPH free radicals is complete (Fig. 1), should be a reference figure for *in vivo* studies on the possible radioprotective effect of the ethanolic extracts of propolis (EEP).

N. Krol*, Z. Czuba*, S. Scheller*, J. Gabrys**, S. Grabriec*** and J. Shani****, (1991). **Anti-Oxidant Property of Ethanolic Extract of Propolis (EEP) as Evaluated by Inhibiting the Chemiluminescence Oxidation of Luminol;** Departments of Microbiology*, Histology** and Embryology, Silesian School of Medicine, 41-808 Zabrze-Rokitnica, Poland, Institute of Parasitology***, Polish Academy of Science, University of Southern California****, Los Angeles, California 90033, USA

Summary
Ethanolic extract of propolis (EEP) has remarkable medical properties, including protection of mice against gamma irradiation. Its anti-oxidative effect has been attributed to its radical scavenging ability. This manuscript demonstrates the ability of increasing

amounts of EEP to inhibit luminol-H_2O_2 chemiluminescence *in vitro*, and suggests that its anti-oxidative capacity is partly due to its high content of flavonoids.

Introduction

The anti-oxidative effect of EEP has been demonstrated so far only indirectly, by its protective ability of food products and by its radical scavenging ability. The purpose of the present study is to demonstrate directly the anti-oxidative property of EEP, by inhibiting a typical oxidation reaction of hydrogen peroxide on luminol, as followed up by chemiluminescence. The EEP's inhibitory capacity on luminol oxidation is compared in this study to that of quercetin, a strong anti-oxidative flavonoid present in EEP in significant amounts.

Results and discussion

Figure 1 demonstrates the ability of EEP to inhibit the luminol-H_2O_2 chemiluminescence. EEP 0.13 mg caused over 86 % inhibition of this reaction, but higher EEP doses could not get any higher inhibition. Figure 2 shows for comparison the inhibition obtained by quercetin. This latter inhibition process is gradual, but reaches a plateau already at a quercetin level of 0.04 mg. As the level of inhibition by EEP and quercetin is about the same (86-88 %) - one can conclude that in this test EEP's anti-oxidative effect per mg is about one third of that of quercetin.

Luminol is easily oxidized by hydrogen peroxide, a free radical production reaction accompanied by omission of light. Inhibitors of radical reactions (antioxidants) inhibit the rate and intensity of the emitted chemiluminescence. The extent of the chemiluminescence inhibition can therefore be used to quantitate the anti-oxidative capacity of any given chemical preparation that will react with luminol and H_2O_2 under the same experimental conditions.

Flavonoids were shown to possess an anti-oxidative property, due to their ability to scavenge free radicals. Their anti-oxidative measure was also verified by their ability to inhibit activities of lipo-oxygenase and cyclo-oxygenase. The anti-oxidative effect of EEP is therefore the outcome of these properties. It can be considered as a regulator of free radical concentrations in various pathological conditions.

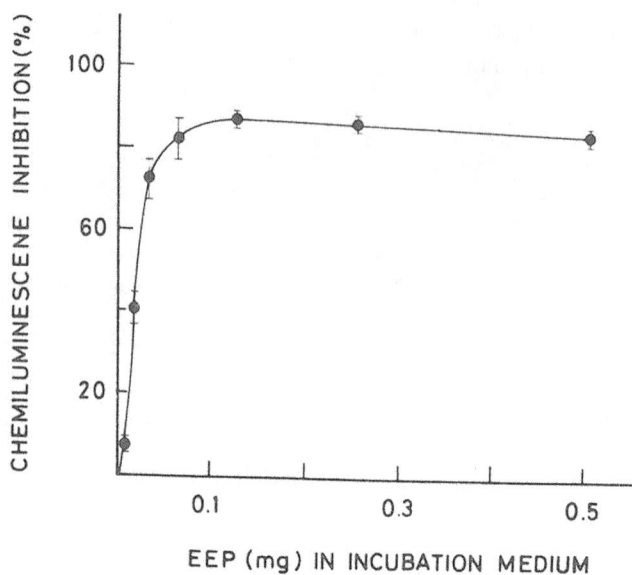

Fig. 1. Effect of various quantities of EEP on inhibition of the chemiluminescence produced by the reaction between H₂O₂ and luminol (Mean± SEM; n=7),

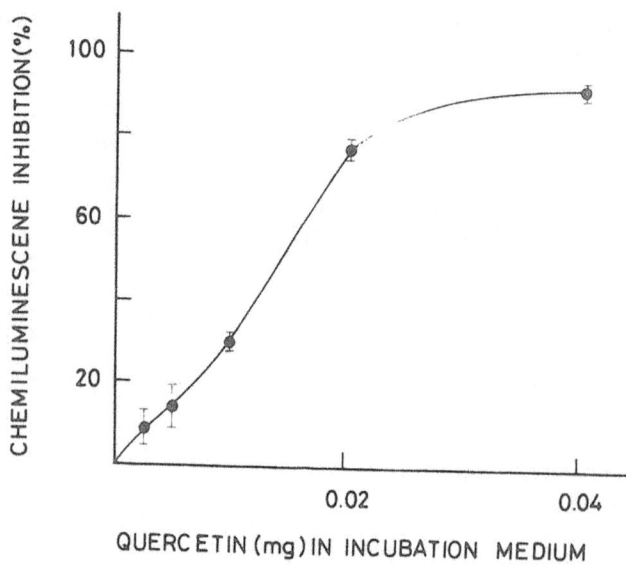

Fig. 2. Effect of various quantities of quercetin on inhibition of the chemiluminescence produced by the reaction between H₂O₂ under the same experimental conditions.

34

Abstracts from **The XXXIInd International Congress of Apiculture** - 1989, Apimondia, Bucharest

281 N.D. Chuhrienko et all. (USSR) - **Complex Treatment of Chronic Bronchitis with Apicultural Products.**

The dynamics of clinical and laboratory data, non-specific resistance and immunoglobulin levels were analyzed in 104 patients with chronic bronchitis. Conventional methods were used in treatment of 56 patients, 48 patients with chronic bronchitis received honey and propolis inhalations in addition to conventional therapy.
The comparison of clinical, laboratory and immunological data gave evidence that honey and propolis are of great effect in the treatment of patients with chronic bronchitis. These patients were discharged from in-patient department 3 - 4 days earlier. The relapse of the disease was 2 times rarer in comparison with a control group.

321 Cora Rosenthal et al. (Israel) - **Demonstration of the Inhibitory Effect of Propolis on Microbial Strains.**

Aqueous and alcohol propolis extracts, obtained from the apiary of Zrifin apicultural station, were studied for antibacterial activity on the examined strains. It was noted that the aqueous solutions showed no inhibition. The alcohol itself exerted a certain inhibitory effect on the growth of certain strains.
The alcoholic extract (30 % p/vol) had an inhibitory effect on the spores of *Bacillus subtilis* and *Bacillus Alvei* and on *Staphylococcus aureus* and *Escherichia coli*, while *Salmonella typhimurium* and *Enterobacty clocea* are practically resistant to the action of propolis. It was noticed that the action of propolis on sensitive strains is of bactericidal nature. The method of groups in agar medium was more interesting than the disc method in proving the inhibition zones. Out of 37 isolated strains in a clinical research laboratory, we have found 2 strains resistant to 6 antibiotics, in general use, which show a remarkable sensitivity to the action of propolis. These latest results motivate our efforts for applying propolis in human therapy.

99 Yang Ruiyu et al. (China) - **The Effects and the Use of Propolis on Veterinary Medicine.**

Scientists and beekeepers abroad have already studied propolis. In 1975, the curative effect on skin diseases and clavus was reported in China. Subsequently there have been experiments on the bacteriostasis of propolis. But there have been no reports on the effect and uses of propolis as veterinary medicine.
In our experiments we tested the effects and proved the usefulness of propolis in domestic animals in combating seventeen pathogenic bacterial species. They are:
Staphylococcus aureus, Streptococcus equi, Erysipelothrix husiopathia, Pasteurella suiseptica, Pasteurella boviseptica, Salmonella abortus-equi, Salmonella cholerae Suis, *Salmonella pullorum, Salmonella gallinarum, Bacillus coli, Listerella monocytogenes, Salmonella paratyphi A, Salmonella typhi, Pseudomonas aeruginosa, Streptococcus B.* and *Sh. flexner.*

35

The XXIXth Congress of Apiculture of Apimondia - 1983 - Apimondia, Bucharest.

C. Ciurcaneanu et al. (Romania). **Treatment of Skin and Genital Herpes and of Herpes Zoster with Aqueous Propolis Extract and Ointment.**

The effect of administration of aqueous propolis extract and of an ointment with propolis in 60 patients with type 1 and 2 recurrent herpes and in 19 patients with herpes zoster are reported on. A control group was treated with classical medicines.
The treatment was efficient in more than 85 % patients with recurrent herpes and over 95 % patients with herpes zoster; the number of pain attacks decreased and the interval between them was longer; no undesirable side-effects were recorded.

Ch. Kalman (Israel). **Apitherapy Success in Israel.**

The author reports good results obtained by the use of bee products in the treatment of various diseases.
With special reference to propolis, the author points out its curative effects in gynecological affections, gangrene and amputations, virus infections (*Asia influenza*), idiopathic thrombosis, severe sinusitis, osteomyelitis, encephalitis viralis, *Escherichia coli*, urinary infections, leukemia and cancer (associated with chemotherapy), which effects are obtained by physicians in Israel.

M. Pavlek-Mocan and B. Briski. (Yugoslavia). **Back to Nature Cosmetics using Propolis.**

The purpose of this work is to point out, for the cosmetic chemist, the problems inherent in the availability of suitable organic compounds or natural bee products, such as propolis with its antibacterial and antiseptic properties. Such products must be checked in order to be utilized safely. They must be non-irritating, non-sensitizing and non-toxic. The authors report on their tests made on cosmetic creams, face lotions, honey bath lotions and honey shampoo with propolis.
The following tests were carried out: 1. Analysis of cosmetic products and determination of propolis; 2. Antibacterial activity; 3. Cosmetic preparations for woman, man and children.

'Propolis' Apimondia 1978, Bucharest.
N. Baidan; N. Oita; Elena Palos; **Using Propolis in Ophthalmology.**

Thanks to its complex chemical composition propolis has a wide range of biological properties useful in medical therapeutics (anti microbial, cicatrizing, antiseptic, antimycotic, antivirus, anaesthetic, trophic properties; it was possible to use it in treating ocular therapeutics after discovering an appropriate solvent tolerated by the eye: ethylene diamine.
Ophthalmic solutions with propolis (2 to 5 %) and propolis ointments (5 to 10 %) were used with excellent results in treating burns and injuries to the eye and also when applied to surrounding tissues: to combat microbial and viral inflammatory infections of the eye

and surrounding tissues in order to maintain asepsis of the eyeball both pre- and post operatively. The ophthalmic solution was rendered considerably more tolerable with the addition of 10 % macrodex as an excipient instead of distilled water.

Propolis processed to constitute a dry (lyophilized) eye wash can be preserved for a long period (at least 1 year), and used as required by adding 10 % macrodex.

The next output is generated from **Compact Cambridge: MED-LINE 1986 Revised for 1990.**

J. Gabrys; W. Krol; S. Scheller; J. Shani: **Free Amino Acids in Bee Hive Product (Propolis) as identified by gas-liquid Chromatography.**

Propolis is a natural resinous product collected by honey bees and containing, among other biochemical constituents, a variety of free amino acids. Acid extraction and quantification of these amino acids by gas-liquid chromatography reveals that their total concentration in this honey bee product is over 40 % w/w, and that arginine and proline constitutes over 50 % of the crude acid extract. As propolis was shown to stimulate mammalian tissue regeneration, we suggest that the physiological significance of arginine in the propolis product lies in its ability to stimulate mitosis and to enhance protein biosynthesis, and that the biochemical importance of proline in propolis, stems from its capability to promote build-up of collagen and elastin, two essential components in the matrix of connective tissues.

J. Simuth; J. Trnovsky; J. Jelokova: **Inhibition of Bacterial DNA-dependent RNA Polymerase and Restriction Endonuclease by UV-absorbing Components from Propolis.**

Several UV-absorbing substances inhibiting the DNA-dependent RNA polymerase of *Escherichia coli* and *Streptomyces aureofaciens*, as well as the restriction endonuclease Eco R1 have been isolated from the water-soluble extract of propolis by two dimensional paper chromatography. The inhibition of bacterial RNA polymerase by the components of propolis was probably due to the loss of their ability to bind to DNA. The general characteristic of the UV-absorbing component of propolis with the most pronounced inhibitory effect upon transcription in vitro is described.

G. Martinez Silveira; A. Gou Gudoy; R. Ona Torriente; Palmer Ortiz MC; Falcon Cuellar **Preliminary Study of the Effects of Propolis in the Treatment of Chronic Gingivitis and Oral Ulceration.** [in Spanish]

For many years it has been known that propolis has a therapeutic value in the treatment of various types of lesion, mainly because it contains more then 30 biological active elements which have so far been isolated and identifiwed. There is worldwide except-ance of the fact that propolis is one of the most useful substances made by bees. Despite of this, its use in our country is of relatively recent date. Several investigators, especially in Matanza, are studying results obtained by the use of propolis for both humans and animals. In other countries it has been used in stomatology, but we have no knowledge of its therapeutic effects in periodontal disease or its effectiveness in th treatment of 37 ulcers of the mouth. Propolan, the name provisionally given to a propolis based substance, has been elaborated for the treatment of chronic gingivitis and types of stomatitis of different etiology. This article presents clinical cases supporting the effectiviness of the treatment as applied by the authors.

C. Myares; I. Hollands; C. Castaneda; T. Gonzalez.; T. Fragoso; R. Curras; C. Soria; **Clinical Trial with a Preparation Based on Propolis 'Propolisina' in Human Giardiasis.** (Cuba) [in Spanish].

The results of a clinical test of an extract of propolis (bee glue) or 'Propolisina' are presented in order to show the effectiveniss of the preparation in treating giardiasis. A study was made of 138 patients, (48 children and 90 adults), divided into two random groups, one being treated with 'Propolisana', the other with an amidazole derivate (tinidazole). The diagnostic method used for the children was duodenal aspiration while in the adults duodenal mucosa frotis was performed by means of duodenoscopy. Similar studies were carried out as a criterion of cure over a 5-day period after the treatment has been completed. Propolisana was used in different concentrations: in children (10 % concentration) there was a 52 % cure rate. In 40 adults (20 % concentration) the results were similar to those obtained with tinidazole; when the propolisana concentration was raised to 30 % there was an improvement in the remaining 50 patients (60 % cure rate in contrast to 40 % with tinidazole). This work shows the success of this natural product, which is easyly to obtained in Cuba. Propolisana showed side effects in the treatment of this intestinal parasitism.

Abstracts from **The XXIXth International Congress of Apiculture of Apimondia**, Bucharest, 1983

N. Varachiu; Cristina Mateescu; N. Luca; F. Popescu; Gh. Pirvuh: **Experimental and Clinical Studies Concerning the Treatment of Several Periodontal Diseases with Apitherapeutic Products.**

The encouraging results obtained by the authors in the period before extraction was intended well as the regeneration of the alveolar conjunctive tissue, periodontal regeneration, action of the gingival mucous membrane, following the application of a propolis and royal jelly based product - Gingiprop - suggested a working hypothesis for some experimental and clinical studies with a new product based on bee venom associated with propolis and pollen, used in the treatment of several forms of periodontal diseases. Within this group of diseases extensively studied by the Romanian school of periodontology we dealt with the lateral abscesses produced experimentally.

S. Roman: C. Mateescu; E. Palos; **Treatment of Some Gynecological Diseases with Apitherapeutics.** (International Congress of Apiculture, 1989)

Ulcero erosive lesions of the cervix uteri, dystrophic disorders, as well as specific and non-specific acute or chronic vaginitis are of major importance in the pathology of the
38 female genital system. They cause pain, lead to discomfort during intercourse, diminish

fertility and favour the incidence of malign metaplasia.

Treatment is difficult and usually long-lasting. Patients dislike electrical or chemical cauterization (searing). Infections and suchlike recur. These were all factors which persuaded some gynacologists tu use propolis in topical application in tincture and ointment form.

Although propolis is known to heal infected wounds and stimulate cicatrization, its scientific use dates only to 1950 when an analyses was started to determine its active components, which include flavonoids, aromatic acids and other elements with antiseptic, antibiotic, antifungal, biostimulative, keratolytic, cicatrizing and even anaesthetic properties already noted by the physicians in their practice.

The use of propolis in gynecology is quite recent.

This study concerns a number of 57 patients with ages ranging between 21 - 58 years, with ulcero-erosive cervicitis, variable in size especially on the exo-cervix accompanied or not by Trichomonas or Candida infections, as well as dystrophic and inflammatory lesions of the vaginal mucus membrane and of the vulva, many of them being chronically infected. This study aims at offering an improved method of treatment.

In addition to the propolis tincture, the propolis spray and the propolis ointment, ovules, (containing propolis, pollen, royal jelly and honey in variable proportions), were used in order to obtain a contact or a quasi- permanent action of these natural products on the affected areas.

Naturally, in specific cervicitis infected with Trichomonas and Candida it is necessary to associate the specific antiparasitic and antimycotic allopathic indication in order to eliminate the pathologic agent, as well as the peroral administration of propolis tablets, these hive products potentiating through the properties of their specific actions. They offer a more rapid cicatrization and regeneration of the cervix and vagina. With the aid of the above mentioned medication very good results were obtained with 42 patients (76.3 %), satisfactory with 9 (16.4 %) and no results with 3 patients. 3 patients were excluded from the treatment because significant symptoms occurred of local allergy (acute pains, vaginal itching, intensification of the congestion and oedema of the vaginal mucous membrane) and the occurrence of the vulvar and peri vulvar erythema with intensive pruritus were noted after a 2 - 3 day period of treatment.

The routine clinical, and the microscopic cytobacterial and if necessary the colposcopic examinations were applied to 54 patients under treatment who were divided into 5 groups according to the lesions, the etiologic agent and the medication applied.

Group 1 - 17 cases of ulcero-erosive cervicitis (12 erosions and 7 ulcerations) with moderate cervicitis with normal microbial flora, prevailing *Staphylococcus* and *Döderlein bacillus* etc. 12 of the patients were previously submitted to various treatments, 3 of them were also submitted to electrical searing. The treatment consisted of paintings on the ulcered areas with propolis tincture or propolis spray once at 2-day intervals and the daily vaginal rinsing with camomile infusion, after which an ovule covered with propolis ointment was applied. The cicatrization and the decongestion of the vaginal mucous membrane occurred between the 13th and the 21st day but the use of ovules was continued for 15 days.

Group 2 - 23 cases of ulcero-erosive lesions of Trichomonas origin among which 16 were not efficient, either through incomplete application or due to the non-participating partner. 4 of these patients were also electrically seared, one of them showing repeated 39 haemorrhages of the cauterized area. The antitrychomonas treatment was repeated, both the topical and the oral one, either for a 10 day period or 2 periods of 7 days with a 6 day period of break.

The concomitant apitherapeutic treatment was similar to the one applied to the first group, recommending the local message of the vaginal mucous membrane with propolis ointment to 14 patients with intensive cervicitis. The ovules were applied alternatively with those of Metrodinazol or Canesten at 12 hour intervals. After a 21 - 35 day period of treatment very good results were obtained in 16 cases, satisfactory with 6 cases and one failure.

Group 3 - 7 cases of ulcerations and dystrophic lesions produced by Trichomonas and Candida, all of them previously treated incompletely and an electrical searing. These patients were submitted to the intensive chemotherapeutical antiparasitic treatment for 10 days, concomitantly with the antifungal treatment lasting 12 - 20 days. The anti therapeutic treatment consisted initially only of paintings or spreading on the affected areas, with tincture or propolis spray and propolis tablets administered orally, also with ovules and the vaginal mucous membrane with propolis ointment applied after the allopathic treatment.

The complex treatment applied was lasting longer than with the first 2 groups, about 35 - 38 days, when 3 very good results, 2 satisfactory and 2 failures were noted.

Group 4 - 4 cases of ulceration and dystrophic disorders of the mucus membrane caused by infections with Candida, among which 3 were treated for a long period with large spectrum antibiotics. The application from the beginning of a complex antifungal, topical apitherapeutic and oral treatment based on the scheme applied to the 3d group, 3 very good results and a satisfactory one were obtained after 23 - 32 days of treatment.

Group 5 - Includes 3 patients at climacterium (age between 55 - 58 years) with hyperkeratosis dystrophy of the vaginal and vulvar mucus membrane and moderate dysuric disorders. The topical application of the above mentioned apitherapeutic products - in cures of 15 - 20 days per month when the regeneration of the vaginal mucous membrane to a suitable smoothness and the disappearance of the urinary disorders were noted.

The following data should be mentioned:
- for the vaginal rinses applied daily or every two days the alternative use of camomile infusion and of bicarbonate solution (and iodine solutions in Candida cases) was applied;
- the treatment with antitrychomonas medicines (Metronidazol, Flagyl, Fasigyn, Canesten) was associated with Stamycine, Micostatin, complex of B vitamins and a suitable diet in order to prevent the occurrence of oral and vaginal candidiasis - incidently noted following the administration of antitrychomonas;
- no matter of the applied method of treatment, and the period of application with allopathic medication in the infections with Trichomonas and Candida, the apitherapeutic medication, much better tolerated, was administered up to the healing and the normal trophicity of the cervix and the vaginal mucous membrane;
- it should be mentioned also that with the great majority of the patients within the first four groups, with which also some disorders connected to the uterus were noted, healing

or clear (obvious) improvements were obtained;
- the interpretation of the good and very good results means the cassation of the parasitic or mycotic infections, the cicatrization of the ulcered erosions through physiologic metaplasia and the regeneration of the trophicity of the vaginal mucous membrane with the normal occurrence of intercourse;

We consider as satisfactory the results obtained with the patients with which the following aspects has been noted:
- the disappearance of the etiologic agents (Trichomonas and Candida);
- the cicatrization of the erosions and the ulcerations, but the mucous membrane was still thick, with the presence of vesiculas or microcysts with haemorrhages and the persistence of the mucous secretion (more abundant) and a moderate grade of discomfort during the intercourse.

Conclusions:
1. Due to its large therapeutical spectrum of properties, propolis is an important remedy in the treatment of the ulcero-erosive and dystrophic diseases of the cervix uteri and of the vagina.

2. Propolis topical application (e.g.: ointments, tincture, spray and ovules) is very important because it provides a quasipermanent contact with the affected areas; when Trichomonas and Candida infections are associated, the obligatory association with the specific allopathic medication potentiate the antiseptic and antiparasitic effect and offers an improvement of the cicatrization and of the period of healing.

3. Although generally well tolerated by the organism the first days of propolis application resulted in the occurrence of some local allergic reactions with 5 - 7 % of the cases.

Excerpts from **Journal of the Royal Society of Medicine, Volume 83, March 1990, p. 159-160**

J. M. Grange MSc MD, R. W. Davey MFHom (Department of Microbiology, National Heart & Lung Institute, Dovehouse Street, London SW3 6LY), **Antibacterial Properties of Propolis (Bee Glue).**

Summary

Propolis (bee glue) was found to have antibacterial properties against a range of commonly encountered cocci and Gram-positive rods, including the human tubercle bacillus, but only limited activity against Gram-negative bacilli. These findings confirm previous reports of antimicrobial properties of this material, possibly attributable to its high flavonoid content.

Introduction
The therapeutic potential of honey has recently been reviewed by Zumla and Lulat. Other bee products, royal jelly and propolis, have also been widely used in 'folklore medicine' for centuries. Propolis is a hard resinous material derived by bees from plant juices. It contains pollen, resins and waxes and large amounts of flavonoids which are benzol-y- 41

prone derivatives found in all photosynthesizing cells. Flavonoids have many biological effects in animal systems but have received relatively little attention from pharmacologists.

We are currently undertaking a screening study of a large number of plants and plant-derived materials in a search for possible new antimicrobial agents, particularly for use against methicillin resistant *Staphylococcus aureus* (MRSA). In this paper we report our findings with propolis and review the literature, mostly from Eastern Europe, on the antimicrobial and other properties of this substance and of its therapeutic applications.

Results

In screening studies at a dilution of 1:20 (i.e. 3 mg of solid material per ml) in nutrient agar, the preparation of propolis completely inhibited the growth of *Staphylococcus aureus* (including the MRSA strains), *Staph. epidermis*, *Enterococcus* spp., *Corynebacterium* spp., *Branhamella catarrhalis* and *Bacillus cereus*. It partially inhibited growth of *Pseudomonas aeruginosa* and *Escherichia coli* but had no effect on *Klebsiella pneumoniae*. Thus it appeared to have a preferential inhibitory effect on cocci and Gram-positive rods. Tube dilution studies showed that it was bactericidal for *B. cereus* and the Gram-positive cocci at dilutions of 1:160 to 1:320, and that growth of the H37Rv reference strain of *Mycobacterium tuberculosis* was totally inhibited at 1:320 and partially inhibited at 1:640.

Discussion

Unknown to us at the time of our studies, the antimicrobial properties of propolis have been well documented in a series of publications from Eastern Europe. Thus it has been shown previously that propolis is more active on Gram-positive than on Gram-negative bacteria. On the other hand, *Listeria monocytogenes* is resistant to propolis which has therefore been used to develop a selective medium for this bacterium. Alcoholic extracts of propolis are active against a wide range of dermatophytes at concentrations of 0.25 to 2 %, antiviral properties have also been described and the protozoa *Toxoplasma gondii* and *Trichomonas vaginalis* were killed within 24 h when incubated with 150 µg/ml of propolis.

The nature of the antimicrobial components of propolis has not been elucidated although there is evidence that they are to be found amongst the flavonoids and various esters of caffeic acid. Caffeic acid phenetyl ester (CAPE) extracted from propolis has also been shown to be toxic for a range of tumour-derived cell lines. A component active against *Bacillus subtilis* has been identified as 3,5,7-trihydroxyflavone (galangin). On the other hand, it has been suggested that the killing of Staphylococci is the result of the combined action of several components, none of which alone are effective. Bio autograms, i.e. chromatograms overlaid with bacteria or fungi in agar media, have revealed that propolis contains more than one agent active against bacteria and *Candida albicans*. The mode of action likewise requires clarification: an unidentified water soluble, u.v. absorbing component of propolis has been shown to inhibit bacterial DNA-dependant RNA polymerases. In addition, synergy between propolis and a range of antibiotics has been demonstrated in several studies. In our studies with the Oxford strain of *Staph. aureus*, we have demonstrated synergy between propolis and an ethanolic extract of *Aralia racemosa*, another plant with antistaphylococcal activity.

Honey has been used as a dressing to promote wound healing. Likewise, ethanol extracts of propolis have been shown to promote the regeneration of bone, cartilage and dental pulp. This may also be a property of the flavonoids which have been shown to be anti-inflammatory and able to stimulate the formation of collagen.

Extracts of propolis are non-toxic in experimental animals. Aqueous solutions (0.5 - 1 %) have been administered to human beings as aerosols for the apparently successful treatment of acute and chronic respiratory disease and have been used as eyedrops. A 10 % alcohol solution has been used for disinfection of hands in dental surgical practice. It appears likely that the beneficial effects of propolis and honey are the result of their flavonoid content and both these natural compounds, and purified flavonoids, appear to be worthy of further appraisals of their therapeutic efficacy.

Bee venom

Since the Middle Ages it has been known that bee venom can be used to combat rheumatic diseases. In the 1930s researchers succeeded in synthesizing a substance resembling bee venom which was used in the treatment of rheumatism. But that substance was rejected because it failed to equal the effect obtained with the natural product. More recent research has focused on analyzing bee venom. New constituent elements have been discovered and special therapeutic applications developed.

Although bees are totally dependent on plants, their venom contains no vegetable substances. Special glands in their abdomen secrete substances which form the mysterious bee venom once they are mixed. If the venom enters the human body for the first time via a bee sting, it can cause considerable swelling and skin irritation. Which is why people have an almost instinctive fear of approaching a honey bee colony.

Bee venom is a powerful substance, more toxic than wasp venom. It can be extremely dangerous for people allergic to it or for those with - for instance - cardiac problems. Given in small doses it can very quickly lead to immunity so that it no longer constitutes a danger. Countless experienced beekeepers have absolutely no fear of their bees and scarcely suffer any ill effects from a few stings. But they always wear a veil to cover their face and protect the eyes. Anyone allergic to bee venom can be made immune by a doctor with the required experience. He will administer injections of venom in steadily increasing dosages which reach amounts equivalent to 2 or 3 bee stings at a time. Once the patient can stand that sort of treatment, sufficient immunity has been achieved.

The active ingredients in bee venom are: apamine, melittin, and approximately 10 phospholipase and hyaluronidase groups, inhibitors which also stimulate the heart and the cortico-suprarenal glands. Two amino acids, rich in sulpho methionine and cystine, are also present. These compounds have an established reputatian in the treatment of rheumatism. The ability of bee venom to stimulate the body's production of cortisone makes it suitable for the treatment of rheumatic diseases, especially arthritis. Directly administered via bee stings or intravenous injections, bee venom is today a well known treatment - sometimes in conjunction with cortisone - in hormone therapy, electrotherapy or apitherapy as practiced in a number of countries (Dr J. Saine, Canada).

Nowadays bee venom is collected on a large scale. A method has been developed of stimulating bees for several hours on their return to the hive by administering light electric shocks at the hive entrance. The workers become irritated and use their stings to pierce a thin film so that the venom sticks onto a glass plate below and the stinging bee is allowed to draw its sting out of the thin layer. The plate is removed by the beekeeper every day. The production per colony is very small, probably a few grams per hive per season.

Bee venom is added to various pharmaceutical preparations. Recent studies have shown that there are many hitherto unknown substances in bee venom (hyaluronidase, phosphatase A, methionine, cystine, mineral salts, etc.). A few principles of how these substances work have been established, but a great deal more research is required.

In the knowledge that beekeepers seldom suffer from any form of rheumatism, and when they do only in a very mild form, a great deal of attention has been paid to the way bee venom combats rheumatism. The venom is administered in various ways:

a) by means of bee stings;
b) by injection;
c) local application and massage with ointment containing bee venom;
d) by administering bee venom at the nerve points also used in acupuncture.

In general it is assumed that direct administration of fresh bee venom via bee stings is the best method. But even when it is mixed into an ointment which ensures that it is rapidly absorbed into the tissue, administration of the venom often leads to very good results with, for example, arthritis.

At present it is possible to obtain an ointment consisting entirely of natural products (see page 40) and which contains substances that stimulate the blood circulation: camphor, hyoskiami and capsici, dissolved in vegetable oil can easily penetrate the skin, carrying with them the bee venom into the tissue. This preparation can be used for all rheumatic conditions, both at the acute and the chronic stage. It can also be used to remarkable effect when muscles have been overstrained in sports and such activities as ballet dancing. Far-reaching improvements can be expected in rheumatic muscular complaints and neurogenic disease.

Work has been carried out on this therapy for scores of years and there is now certainty that the treatment not only provides symptomatic relief but also brings about long-lasting cures.

The painful areas must be subjected to repeated treatment, even when the pain has disappeared. Especially in cases of chronic rheumatism ample time (1 to 2 years) can be required before a total cure is achieved.

Treatment

The ointment should be applied directly to the painful area in a layer 1 to 2 mm thick. In order to keep the ointment active a damp cloth should be laid on it. If the place being treated is suitable, a further nylon cloth can be wrapped around it with flannel on top of that. The treatment should be applied mornings and evenings. It has been shown that if the acupuncture points along the nerve path leading to the site of the pain also have a small quantity of ointment rubbed into them, a cure is achieved more rapidly.

Because the ointment contains bee venom it may not be used on the mucous membranes of the mouth, nor on the eyes or genitals. In order to avoid getting even tiny quantities of the ointment on sensitive areas, the hands should be thoroughly washed after application of the ointment

K. Rader, A. Wildfeuer, F. Wintersberger, P. Bossinger and H.-W. Mucke.
Characterization of Bee Venom and its main Components by High-Performance Liquid Chromatography. (1987).
(Research and Development, Heinrich Mack Nachfolger, D-7918 Illertissen, F.R.G.) In Journal of Chromatography 408, 341-348.

The main components of bee venom are the enzymes phospholipase A and hyaluronidase and the low-molecular-weight proteins melittine and apamine. Many methods for the characterization of bee venom have already been described. These either measure the biological effects of total bee venom or those of its individual components or employ typical chemical reactions of the proteins or enzymes. Bee venom exhibits bacteriostatic and bactericidal properties[1,2] and a slight direct hemolytic effect[3]. Other typical characteristics are hyaluronidase activity[4] and the formation of lysolecithin by phospholipase A, which enhances haemolysis[5]. The delay of the coagulation of a heated egg yolk suspension[6] and the formation of CO_2 by enzymic cleavage of phospholipids[7] are other typical effects caused by phospholipase A. From the chemical point of view, bee venom shows of course all typical protein reactions and can also be separated by means of electrophoretic and chromatographic procedures[8].

Conclusions

We have developed two HPLC methods with which bee venom can be determined qualitatively and quantitatively, if external standards of bee venom or of the individual bee venom components are available. With regard to precision and handling, the method is superior to the previous reported column chromatographic methods. Acceptable separation results were only obtained by using exclusion chromatography, where the components of the bee venom elude according to their molecular size. Adequate separation of apamine, a minor component, required a subsequent reversed-phase chromatographic step.

From the biological point of view it is interesting that the bee venom from various sources showed no appreciable differences in composition. From the analytical point of view this seems to allow a standardization of bee venom with respect to the proportions and amounts of the components.

References

1. S. Ortel and F. Markwardt, *Pharmazie*, 38 (1955) 743.
2. A.W. Benton, R.A. Morse and F.V. Korsikowski, *Nature (London)*, 168 (1963) 295.
3. W. Neumann and E. Habermann, *Arch. Exp. Pathol. Pharmakol.*, 222 (1954) 367.
4. E. Habermann, *Biochem. Z.*, 329 (1957) 1.
5. E. Habermann and W.P. Neumann, *Biochem.Z.*, 328 (1957) 466.
6. E. Habermann and K.L. Hardt, *Biochem*, ,50 (1972) 163.
7. E. Habermann, *Biochem.Z.*, 328 (1957) 474.
8. E. Habermann and K.G. Reiz, *Biochem.Z.*, 341 (1965) 451.

Abstracts from **The XXIXth International Congress of Apiculture of Apimondia,** Bucharest, 1983

Elena Palos; Filofteia Popescu; **Use of Bee Venom in Antirheumatic Drugs. (Romania)**

Bee venom is a biological product, with a complex composition of non-volatile substances (enzymes, proteins, amino acids), as aqueous solution, and volatile organic compounds (30 - 60 %) of its weight. Immediately after its collection, bee venom is a colourless liquid, clear, with a sweet taste, a little bitter, astringent, with a smell producing 47 causticity on the mucous membranes. It is water soluble, insoluble in alcohol and ammonium sulfate. In contact with the air it forms, opaque, white or grayish-white crystals, depending on the season in which it was collected.

Below 0° C. conservation is good; protected from moisture it keeps its properties and therapeutical value for a long time. Bee venom is a unique complex substance with its own characteristics.

It differs from snake (viper) venom, which is coagulant, while bee venom is hemorrhagic. Its components: apamine, melittin, phospholipase, hyaluronidase, show two different actions: first of all they are inhibitors of the nervous system, and secondly they act as stimulators of the heart and the adrenal glands.

Bee venom contains also mineral substances, volatile organic acids, formic acid, hydrochloric acid, ortho-phosphoric acid. Some antibiotics, an enzyme - phospholipase A, as well as two amino acids rich in sulfur methionine and cystine are also present. The latter two compounds have an outstanding action in the control of rheumatism. It has been proved that sulphur is the main element inducing the release of cortisol from the adrenal glands, a basic element against infections of the organism. Simulating the release of cortisone (cortisol), bee venom acts, as already known, in the treatment of rheumatic diseases and especially arthritis. Applied directly or by intramuscular injections, alone or associated with corticoids, bee venom is recognized and used actually as an active factor in the treatment of such diseases.

Combined with the other usual treatments, electrotherapy, hormone therapy, eliminates the periarticular deformations, improves the joints movements and the general state of the organism.

An ointment - Apireven - and a liniment are developed in Bucharest. Both products were applied in several cases of rheumatoid polyarthritis showing important muscular participation and diminution of the pains in sciatic, hyperalgic neuralgias, neuromial-gias, intercostal and bronchial neuralgias. An increase of the cutaneous temperature with change of the blood circulation were noted with several patients previously shown to be suffering from phlebitis and thrombophlebitis.

Good results were also obtained in all cases of sciatic neuralgias and circulatory phenomena in neuromyalgia and articular pains.

With chronic neuro inflammatory affections the results were very good, showing the ceasing of pains and the partial recuperation of the movements.

Under repeated topical application the effect is long-lasting.

Associated with a special oral therapy (vitamin therapy, etc.) good analgesic effects were obtained in cortisone-dependent rheumatoid polyarthritis. 4 - 5 days after topical application a 2 - 3 day period of interruption (break) is necessary, then the treatment is applied again. No local incidents were noted.

Royal Jelly

a concentrated bee product

Royal jelly is the food secreted by the worker bees which is given to the growing queen larvae. The same food is also given to the worker bee larvae for the first three days after they hatch. The queen larvae are fed the jelly throughout the entire period of their development until they pupate. Royal jelly is not only a rich food for the larvae concerned: it also turns out to be a highly valuable substance when examined in the light of human physiological needs.

We now know that royal jelly consists of roughly two-thirds (66.5 %) water and the rest dry matter. The following substances can be found in the percentages given (of the whole, including water): protein, 12.34 %; fats, 6.46 %; sugars, 12.5 %; ash, 0.82 %; unidentified substances, 2.84 %. It contains many vitamins, types of hormone and other revivifying substances. The royal food has an extremely complex chemical structure which imparts to the queen larvae a completely different form. The queen even possesses unique organs and working mechanisms. Over the three days during which they are given royal jelly, the worker larvae undergo a stormy growth process: they become 250 times heavier than the original egg. The queen, fed royal jelly throughout her formative period, is born five days earlier than the workers and is twice as heavy. In addition, she can live for as long as five years and in the most intensive breeding period can produce up to 2,000 eggs daily, considerably more than her own weight. A worker bee lives barely six weeks in the summer, and winter bees do not survive longer than eight months. The effect royal jelly has on the bees is nothing short of miraculous and it would seem clear that very extraordinary substances are responsible for this.

Royal jelly was first recommended for human therapy by Professor Chauvin in 1922. In France particularly intensive studies were made and later a monograph was published: B. de Belfever: *La gelée royale des abeilles;* (ed. Librairie Maloine, Paris 1958). The substance was prescribed for hospital patients under strict scientific control.

Different medicall specialists have carried out tests. Research focused mainly on the effect of this natural product on neurasthenia, stress, convalescence, combating the symptoms of aging, growth problems and pregnancy.

Royal jelly is also a superb ointment for the skin. It feeds and regenerates tissue and is extremely helpful in removing uncomfortable or itching scars. When used in face creams, it produces truly miraculous results.

Abstract from **The XXXIInd International Congress of Apiculture,** 1989, Bucharest

Shi Bolun, et al.; **Clinical Observation on Curative Effect of Freeze Dried Royal Jelly Products on Hyperlipemia and Diabetes.** (China)

Patients with hyperlipaemia or diabetes identified through clinical tests were given freeze dried (lyophilized) royal jelly products at a dosage of 1 gr per day, b.i.d. on an empty stomach. They were put on a normal diet and were taken of lipid-reducing drugs. From July through December 1988, 60 cases of hyperlipaemia and 12 cases of diabetes were observed. The freeze dried royal jelly products were administered for period ranging from 1.5 - 5 months. Blood sugar and lipid levels wre tested before and after administration.

Pollen

Pollen can be collected and processed for therapeutic use in two ways. The usual method is for the beekeeper to fit a pollen trap at the entrance to the hive in order to capture the clumps of pollen attached to the bees' hind legs. The compact mass formed by the pollen grains offers some protection against oxidation caused by atmospheric oxygen.

The second method consists in collecting what is known as bee bread. This is pollen stored in the comb and is better provided with animal ferments. It is first chewed by young bees, the fermenting agents are added and it is then mixed with honey and stirred around the brood area on the comb. The beekeeper can remove these combs from the hive, cut out the parts of the comb carrying the bee bread, freeze it and then grind it to powder. This is the best type of pollen for internal use, but it is difficult to obtain and much more costly.

Clumped pollen can also be ground and used in powder form but it then has a shorter shelf life. For some considerable time there was a certain amount of doubt as to whether the pollen could be absorbed by the stomach since the minute pollen grains are covered with an indigestible waxy substance. Attempts were made to circumvent the problem by grinding and fermenting the pollen. But Kerkvliet (1985) has demonstrated that each pollen grain has tiny openings in the external membrane which allow the digestive juices to penetrate. The pollen grains are thus completely rinsed out. Grinding and fermenting pollen is not advisable since these two processes shorten its shelf life. The adverse effects produced by grinding the pollen can, however, be avoided if the pollen powder is mixed with honey immediately after grinding, the honey acting as a preservative.

Pollen is best for internal use when it shows a wide spectrum of colours. This means that the pollen clumps should be clearly of different colours, giving proof that the pollen comes from various 51 species of plants, something which improves its effectiveness. The beekeeper or supplier of the pollen can usually achieve this by mixing various kinds of pollen, even pollen from various locations where necessary, and to use pollen gathered throughout an entire season so that a maximum of different plant species are repre-sented.

The following information regarding useful substances in pollen comes mainly from 'Heilwerte aus dem Bienenvolk' (Herold, 1970) and 'Les vertus merveilleuse du pollen' (Caillas).

Pollen contains many substances of value to our health. Even more important than their mere presence is the balanced mutual interaction between the components. Since bees visit so many types of plants, pollen can be multi-coloured as well as extremely variable in composition. In general, however, it contains on average 5 - 25 % proteins and 0,8 - 15 % oils, approximately 40 % of the latter consisting of amino acids or unsaturated fatty acids. Of the 22 amino acids known at present to biochemistry, 20 have been found in pollen. There are proportionately more of the following amino acids in pollen than - for instance - in meat, eggs or cheese: cystine, histidine, arginine, isoleucine, leucine, lysine, meth-ionine, phenylanaline, threonine, tryptophan, valine and glu-conic acid. In addition pollen contains approximately 40 % carbo-hydrates, vitamins, antibiotic substances and growth-promoting substances. Quite probably the often spectacular results obtained with the use of pollen can be ascribed to the proteins, the vitamins and the amino acids, though the mineral trace elements should not be forgotten either: the presence has been detected of iron, copper, potassium, sodium, magnesium, calcium and silica.

Vitamins

The following vitamins are present in pollen: provitamin A and carotene, which the organism uses to synthesize vitamin A; B1 (thiamine); B2 (riboflavin or lactoflavin); B5 (pantothenic acid), which promotes growth, skin care, strengthens the nerves, im-proves breathing, regulates the digestion, improves the blood; B3 (niacine) and PP (PP is a nicotinic acid and is well known as a substance which protects against pellagra) which counteract disorders in balance, are important for the skin, metabolism and the nerves; B6 (pyroxidine) which stimulates growth and combats anaemia; vitamin C; folic acid 2 complex, important for the renewal of blood cells; vitamin E (tocopherol) which improves fertility and conception by its effect on the production of sperm and ova; vitamin H (biotin), a growth stimulant which is also effective against skin infections; rutin strengthens the capillaries thus preventing undesirable internal bleeding. Rutin also strengthens the heart muscle. Finally, pollen contains antibiotic substances. This has been shown from research in France performed by Chauvin (1959) and Lenormand.

Caillas recommends a pollen cure in cases of disturbed metabolism. Pollen has a curative influence on microbes, bacteria and fungi in the stomach and bowels - on the digestive flora. Pollen regulates the bowel function.

Growth-promoting substance

The growth stimulant present in pollen has so far only been tested on mice and rats. Animals fed on pollen show remarkable improvements in growth.

Effective components

Prof. N. Joirisch (1969, USSR) writes that pollen contains everything required by the human body. Since it is derived from so many different flowers, it contains effective substances from many plant species. The various types of pollen have the following ingredients in common: protein, amino acids, sugars, minerals, vitamins, hormonal substances, fats and aromatic substances.

Pollen contains practically all the known amino acids in great quantities. Approximately 200 grams per day is sufficient to provide the daily amino acid requirements of an adult. Study of pollen ash has revealed the following minerals: 30 % potassium, 7 % magnesium, 8 % calcium; in addition, 7 % iron, 0.5 % chromium, 7 % phosphor and minute quantities of other minerals. (J.D. Kerkvliet, 1989)

Pollen contains minerals and vitamins. The quantity depends much on the kind of blossoms visited by the bees. High-desert 53 honeybee pollens, for example, contain the following values, (published as chemical analyses by the 'High-Desert Honeybee pollens' Company; values stated as mg/100 gr); compared to the average values in Europe (in brackets)

MINERALS.
Calcium: 151 (321), Iron: 7.5 (17), Potassium: 566 (1329), Phosphorus: 434.5 (286), Sodium 10 (161), Magnesium 99 (207), Zinc: 5 (12.7), Copper: 0.79 (1.1), Manganese: 5 (4.5)

VITAMINS.
A - Beta carotene: 0.44 (11), B1 (Thiamine): 0.71 (1.02), B2 (Riboflavin): 1.64 (1.53), B3 (Niacin): 9.11 (9.26), B6 (Pyridoxine): 0.43 (0.32), Biotin: 0.007 (2.5), Folic acid: 0.72 (0.96).

These values demonstrate that the quantity of minerals and vitamins vary greatly between different types of pollen. It is advisable to mix several types of pollen to ensure that as many organic materials are gathered as possible. The high-desert type of pollen is clearly gathered (as analytical research proves) from a monoculture of cactus plants.

Indications
The literature ascribes the following results to the use of pollen:

1. Pollen stimulates the appetite.
2. Pollen is used for digestive disorders.
3. Pollen has a positive effect on the psyche.
4. Pollen can be used in nervous debility and other complaints originating from the nerves.
5. Pollen improves the general health, reinforces the body and provides improved resistance to illness.
6. Pollen improves blood circulation and improves the thought processes.
7. Pollen strengthens the coronary arteries and has a beneficial effect on cardiac muscle tissue.
8. Pollen stimulates growth.
9. Pollen improves vision.

10. Pollen promotes growth of the hair.
11. Pollen is an effective cosmetic. Mixed with egg-white it can be massaged into the skin and left to work for half an hour; it refreshes the skin and improves skin tone.
12. Pollen is effective in prostate cases. A third of all men in Western Europe have prostate problems: inflammation, enlargement or cancer of the prostate. Every male aged over 45 is recommended to have a medical check-up.

We can safely say that pollen has extremely beneficial effects on the digestion, it stimulates the appetite, it improves conditions of physical weakness and can be useful in the treatment of neuroses, depression, neurasthenia, prostate problems and diabetes by improving resistance to these illnesses and stimulating the endocrine glands.

How much pollen is needed on a daily basis?
A daily dose of 60 grams of pollen is sufficient for all our requirements as regards both vitamins and amino acids. But seeing that all our food contains vitamins, it is not necessary to take so much pollen every day: one tablespoonful per day is sufficient.

In his book Caillas (1945) indicates that 32 grams of pollen per day constitutes a 'blitz' cure. He also states that a daily dose of 15 grams can be taken for long periods.

But in fact a tablespoonful per day works extremely well, improving the condition of the body and building up resistance to infection.

The best way of taking pollen is to chew it slowly and mix it well with saliva. It should be taken on an empty stomach. If the taste is found to be disagreeable, the pollen can be mixed with water, milk or yoghurt (particularly if a little honey is added).

Pollen can also be used with honey on bread. The honey should be spread directly on the bread, the pollen sprinkled on the honey and, if desired, butter over the pollen. An even better idea is to mix propolis or royal jelly with the honey. In fact we should use pollen every day. Even a teaspoonful with our breakfast every morning is sufficient to promote better health.

It is always possible to stop taking pollen but there is no harm in continuing to take it if you are somewhat out of condition. Requirements differ from individual to individual. When you do become ill, try using pollen in combination with other bee products. There are wonderful components in all the substances produced by bees, all of which have a specific effect on the human body.

Bee bread

The bees make bee bread by processing pollen. The pollen is chewed by young bees and packed tightly in the cells. They apply fermenting agents and ensure that pollen deteriorates less quickly. Pollen processed in this way is known as bee bread. The only way it can be prepared for use is to freeze the comb and grind it finely. This means, of course, that beeswax gets into the bee bread, but the wax is an excellent substance. Bee bread has characteristics which are even better for the human organism than the unprocessed pollen clumps collected in a pollen trap before being stored in the cells. But it is difficult and expensive to collect. The beekeeper will not be all that keen to see his combs disappear since it is not easy to build up a store of combs again. Fortunately however, the crude pollen collected as clumps from the bees' legs is a valuable substance in its own right.

Abstracts from **The XXXIInd International Congress of Apiculture** - 1989, Apimondia, Bucharest

Xu Ligen et al. **The Efficacies and Mechanisms of Pollen in Anticancers and Anti-aging.** (China)

The article is a comprehensive report on the effectiveness of pollen in treating cancers and aging; the article also provides information on the mechanisms involved. The authors believe that pollen increases the immune functions of the body, postpones aging and inhibits cancers. The effective components of pollen, such as vitamin A, C, D, E and S and trace elements Se, Fe, Zn, Mn, Cu, I, and F, normal elements Mg, Ca, active biosubstance, enzymes and nucleic acids all play an important role in the above conditions. The pollen serves as a supplementary agent for cancerous patients undergoing chemotherapy and radiotherapy and it promotes health as a contribution to longevity.

Wang Weiyi. **Study on the Digestibility and Absorbtivity of Unbroken-Walled Pollen. (China)**

Can the nutrients of unbroken walled pollen be digested by the human body? What is the difference between the digestibility of broken and unbroken walled pollen? Many scholars show much interest in them and hold different opinions. We experiment with Rhesus monkeys (*Mucaca mulatta*) to test whether they could digest the amino acids, vitamin C and mineral element K of rape pollen and to make a preliminary assessment of the mechanisms involved.

The results of the experiment indicate that the pollen wall has no effect on the digestibility of amino acids, vitamin C and the mineral element potassium in pollen. The mechanism involved in the digestion of unbroken-walled pollen is possible in the following: after the pollen grains reach the digestive tract, the change in osmotic pressure and the cracking of the germ pore result in the digestion of the internal wall by digestive enzymes and in the absorbtion of nutrients. The digestibility of nutrients from other types of pollen by Rhesus needs further research.

Abstracts from **The XXIXth International Congress of Apiculture of Apimondia** - 1983, Bucharest

E. Gheorghieva; V. Vassilev; **Pollen in Treatment of Chronic Liver Lesions.** (Russia)

Pollen was administered to 50 patients, of whom 25 suffered from hepatic steatosis and 25 of chronic hepatitis. Two spoonfuls of pollen were administered daily. Laboratory tests (bilirubin, transaminase, electrophoresis) and clinical examination showed improvement in the condition of the patients thus confirming the therapeutic value of pollen.

E.Gheorghieva; V. Vassilev; **Pollen used in Anaemia.** (Russia)

Pollen was administered to 50 patients with post-hemorrhagic anaemia and showing iron and B12 deficiency. Pollen was well tolerated by patients. The treatment is easy to apply, for both inpatients and outpatients.

Honey

Honey is extremely hygroscopic and can be stored for long periods if (in general) it contains less than 17 % water and is stored in an airtight container. As soon as the water content of honey is increased by the exposure to humidity, it can start to ferment. The bees place honey in the cells and close them off with wax lids so that after the drying process in the air warmed by the bees, moisture can no longer reach the honey. It sometimes happens that the beekeeper processes his honey too early so that it will not crystallize properly. Or the honey can form rough crystals so that water appears on the surface and the upper layer of honey starts to ferment.

It is always preferable to obtain honey from a local beekeeper who processes his own honey (cold processing) rather than to buy it in a shop. Honey from a regular shop is almost always imported, with the disadvantage that it has been packed in barrels in which it has stood for so long that the honey has crystallized. Consequently, before the importer can put it in jars for sale, the honey has to be heated so that it liquefies. Consumer legislation allows honey to be heated to 60° C for as long as 24 hours. This process causes the loss of most of the honey's fermenting agents, representing a major loss of important ingredients.

Sometimes a layer of pollen is left in some cells and it is covered with honey and thus preserved. This means that some pollen will get into the honey, but the amounts are very small.

Composition

Honey sugar content

Honey is a product made from nectar (plant juice from flowers and other sources) dried and supplied with ferments by the bees with a water content of between 20 - 80 %. The bees employ various methods to remove the excess water by evaporation. The plant juices are thus thickened in a natural

way to form honey and, in addition, are provided with fermenting agents by the bees. Average quantities of sugars in honey are
as follows (White, 1962):

fructose or levulose	38.19 %
glucose or dextrose	31.28 %
cane sugar	1.31 %
malt sugar and other double sugars	7.31 %
multiple sugar (dextrine)	1.50 %

Levulose (laevus = left-handed) is known as a levorotary type of sugar because it bends
polarized light to the left.
Dextrose (dexter = right-handed) is known as a dextrorotary type of sugar for the
opposite reason.
Both are single sugars: glucose and fructose together are known as invert sugar, which is
levorotary probably because fructose is more strongly levorotary than glucose is
dextrorotary.

Minerals

Human beings require daily amounts of the minerals potassium,
phosphor, calcium and magnesium (positive) plus chloride and
phosphate (negative). The following quantities of minerals in
milligrams can be found in 100 grams of clover and heather honey
respectively:

potassium oxide	10,30	13,30
phosphor oxide	104,70	204,60
magnesium oxide	3,80	5,90
chloride	19,00	16,70
phosphate	16,,40	15,10
silicic acid	4,40	19,00
iron oxide	0,60	0,60
manganese	0,40	0,90
copper oxide	8,70	14,90

It could be said that every little bit helps, but relatively seen the
amounts are very small. If, for instance, we assume that we have a
daily potassium requirement of 2 grams, by eating 100 grams of
honey per day we absorb approximately 10 mg (0.01 g) of
potassium per day. But we require far smaller quantities of silicic
acid, iron oxide, manganese and copper oxide - and then every
little bit does help.

There are nine minerals which have to be absorbed into the body in tiny trace amounts, namely iron, copper, manganese, cobalt, molybdenum, zinc, silica, iodine and fluoride. Of these iron, copper, manganese and silica are demonstrably present in all types of honey. Pine honey contains more iron than other types (30 mg of iron in 100 g). Copper is found in darker types of honey e.g. heather honey contains 15 mg of copper in every 100 g. Acacia honey contains the most manganese (20 mg per 100 g).

Fermenting agents

In contrast to crystallized sugar - which only consists of sugar, possesses no minerals and is a completely dead substance - honey contains all kinds of fermenting agents or enzymes: these are the catalysts which allow all kinds of nutritious substances to be digested. Natural honey is alive because of the fermenting agents it contains.

A number of these agents are known:

Invertase

This ensures that the saccharose (a double sugar) is separated into single sugars, such as dextrose and fructose. Invert sugar is the largest component of honey. Far from being lost in the transformation process, the invertase enzymes remain active in changing saccharoid molecules.

In the process that transforms nectar to honey, this separation is accelerated because the young bees take up the nectar time and time again in the drying process - on average 50 % of the water content of nectar has to be removed by evaporation - and the worker bees add invertase enzymes to it each time.

Even when honey is stored the inversion process continues. Fresh honey can contain as much as 8 % invert sugar, a proportion which drops to a mere 2 % after a few weeks.

Invertase is the principal fermentation agent in the ripening process of honey. But invertase is extremely sensitive to heat. Even long-lasting heating to 45° C. considerably reduces its effectiveness and it fails to work altogether at a temperature of 60° C. The invertase content of honey can therefore be used to determine whether or not the honey has been subjected to heating.

The law allows imported honey to be heated to a temperature of 60° C. for 24 hours. This means that not only is the enzyme content reduced by half but the invertase is completely destroyed. And yet the legislators accept this procedure in order to meet the requirements submitted by the trade. Hence the importance of obtaining your honey directly from a beekeeper who uses cold extracting methods.

However as we shall see further on, a number of fermenting agents do survive the battle.

There is a danger of honey heating to over 45° C. when it is stored in warehouses in warm countries. Also in such countries it is customary to heat honey to 75° C. in order to kill off any yeasts that may be present. This is particularly important in the case of honey with a high water content. The treatment prevents the honey from fermenting any further. And there you are - a superior type of honey for the consumer's lack of taste: in general the public are not happy with crystallized honey, preferring it to be runny (so a little extra water does no harm) and public taste also requires light-coloured clear honey. The latter is achieved by cleaning the honey by passing it through filters at high pressure. Despite all this, there is still some of the original goodness to be found in most types of honey. But at the risk of sounding repetitive: 'buy your honey directly from a beekeeper employing cold methods of extraction.'

Amylase (Diastase)
There are two amylase fermenting agents which break down starch. Honey contains one that turns starch into dextrine and a second that breaks down maltose. The first is knocked out at 70° C. and the latter at 50C. Each type of honey can have a different enzyme content. Since the pollen content is variable, the enzyme content is also variable, but most of the fermenting agents are added by the bees. If the weather is hot and the air is dry when the nectar is collected, the water content is low. Consequently, the bees have less work to dry it and the process of turning nectar to honey becomes less thorough. In any case, there is little time, since as much as several kilograms of nectar can be collected in a single day.

This sort of honey flow is used by the large honey producers in 61 the US and Australia, which means that their honey is usually less thoroughly fermented.

Glucose oxidase
In the fifties, White discovered the glucose oxidase enzyme, a fermenting agent that has also been isolated from fungi such as penicillin. This substance turned out to remain effective even after being heated in a sealed container. It turns glucose into gluconic acid, the principal organic acid in honey. It is chemically related to glucuronic acid and it binds to substances which are difficult to dissolve so that they can be removed from the kidneys. One of its uses is the removal of toxins from the body.

Fermentation
Fermenting agents which we produce ourselves and those made by the honey bee act mainly during the chewing of food. In the stomach too they can continue for some considerable time to affect the food consumed. To increase the effective working-life of honey enzymes, the honey should be spread directly onto bread (with butter on top of the honey) and it is also possible to allow honey to work on, for instance, some muesli before eating. If honey is used for baking it should be placed in the dough and allowed to work in for a time before it loses most of its effectiveness by being heated.

Hormones
The vegetative nervous system is divided into sympathetic and parasympathetic systems. The sympathetic system is the active one. The parasympathetic system controls and provides energy: it sees, for instance, to the movements of the heart, the arteries and the bowels. Stimuli which set the various organs to work depend - partly in the sympathetic system and wholly in the parasympathetic - on a substance known as acetylcholine, which is also to be found in honey. Acetylcholine can stimulate the heart which enables the body to take up more fuel - in the form of glucose. Which is why honey is sometimes known as the "oats of the heart". Cardiac patients can derive a great deal of profit from honey.

Acetylcholine also improves the metabolism and regulates the blood circulation. It stimulates the digestive organs and the kidneys. Acetylcholine retains its effectiveness even when dissolved to one part in several thousand millions. It is well preserved in honey and is resistant to long periods of heating. A kilogram of honey contains anything from 0.6 to 5 mg of acetylcholine. This suffices toIt assist the liver in the efficient processing of sugars, which are difficult to digest. It is clear that it is much better to use honey instead of granulated sugar.

Inhibines

These are substances which inhibit the growth of bacteria or kill them altogether. Hence the active antibacterial properties of honey. Buchner and Kopp (Freiburg, 1967) discovered that the bactericidal properties of honey remain effective even when the honey is heated in a closed container. This demonstrates that it must contain an essential oil which is largely responsible for the specific bactericidal action.

Honey retains its effectiveness even in a 1: 64 solution. The inhibines cannot originate solely from plants since the bactericidal properties are present even in honey made by bees fed on sugar, though this is less effective than natural honey.

The bees themselves use nectar to sterilize pollen.

Acids

Many types of acid are present in honey as agents producing aroma and taste. They stimulate the appetite and the digestion.

Aromatic substances

Cremer and Rietman(1946) used gas chromatography to demonstrate the presence of 50 types of aromatic substances in honey, 23 of them identified. They consist mainly of aliphatic alcohols, aldehydes and ketones.

Vitamins

Honey contains very few vitamins. Pollen and royal jelly contain many more and in greater quantities. Any vitamins found in honey are there mainly due to the presence of pollen.

Amino acids
There are few of them in honey but they do occur in pollen.

Honey to improve bodily functions

Honey in its natural state is generally used to guard against illness and to improve bodily functions. A few examples:

Pregnancy
In cases of toxaemia of pregnancy intravenous doses of honey (M2, produced by the Woelm company in Eswege, Germany) can be administered. The preparation consists of 40 % honey without proteins. It helps prevent vomiting and improves the appetite. The honey solution protects the liver tissue during bouts of toxaemia. (Royal jelly also provides vomiting and gives the pregnant woman a general feeling of well being.)

Infants and children
Honey is prescribed to encourage growth in premature babies, as dietary aid in dyspepsia (metabolic disorders), in cases of malnutrition and as a supplementary food when breast-feeding is insufficient for the child's needs.

Honey poduces no side effects in babies or infants. Weight gain is stable and greater than average. Fewer infections occur. Honey is easy to prepare and hygienic (it contains its own disinfectants). Intake of honey causes calcium and magnesium binding in the body. The teeth develop better. Honey helps in cases of bed-wetting, helps to calm children down and improves their sleep patterns.

Wounds, oedema, bacteria
Honey has a healing influence on suppurating wounds and on infected tonsils. It promotes rapid epithetialization and absorbtion of oedema from around ulcer margins. It inhibits the growth of bacteria responsible for bowel infections and is also effective in the treatment of typhoid, paratyphoid, peritonitis, dysentery and cholera.

Honey for convalescent patients, for the infirm and for the aged

In 'Physikalische Medizin und Rehabilitation' (November, 1967) Dr S. Kaul published an article about the intravenous doses of honey already referred to (M2, produced by Woelm) with which he treated patients who had just undergone surgery. He also used it for pregnant women, underweight patients and people with general symptoms of exhaustion. The study contained not only subjective statements about improvement in health but also objective observations concerning improved appetite, weight gain, haemoglobin increase, return to normal sleep patterns and stabilization of blood pressure. In the past he used to treat such patients with intravenous glucose. Now he regards the treatment with honey not only more effective but also, in fact, to be preferred. Intravenous glucose leads to an 18 % improvement in glycogen synthesis as compared to ordinary sugar. A solution of honey is much better: it is 68 % more effective than glucose. Lemmpt suspects that this is connected with the discovery made by Prof. Koch (1949) regarding the cholinergic effect mediated through acetylcholine, which stimulates the use of sugars. In addition, honey improves the circulation and coronary blood supply.

Honey stabilizes blood pressure, regulates heartbeat (extra systolic arrhythmias), shortens the cardiac contraction time (systole) and lengthens the relaxation time (diastole). Honey provides reserves for the heart, which can only benefit this vital organ.

From the above it is clear that honey is beneficial for elderly people. In almost every case, older people have something wrong with their circulation, with their metabolism and with other systems. In Russia there are beekeepers - especially in the Caucasus - who have always taken not only honey but also pollen and royal jelly. A remarkable number of them have lived to be more than a hundred years old, the oldest being 140. This would seem to bear witness to the benefits accorded to these old people by the bee products they have ingested.

The liver

The liver produces gall, enzymes, glycogen, blood protein substances (albumen, globulin), clotting agents and uric acids. The 65 liver also plays an important role in supplying colouring agents for the blood. It manufactures glucuronic acid ester and oxidizes toxic substances. It stores nutrients, including 300 grams of the 400 to 600 grams of glycogen (animal-derived starch) present in the body; it contains blood (20 % of what the body contains), water, mineral salts, certain vitamins (e.g. B12 which protects against anaemia) and heavy metals.

The liver is the principal organ in the metabolism of sugars, fats and proteins. It regulates the distribution and delivery of nutrients to the blood, which means that the liver is the main organ regulating body temperature.

The liver therefore works hard and determines all sorts of bodily functions. It is of the utmost importance that we assist this major organ, so that it should suffer as few defects as possible, defects which will affect many other bodily functions. Honey is the only substance that enables us to go a long way to achieving this, since honey assists the liver in three ways:

> by promoting the formation of glycogen
> by removing toxins from the liver
> by stimulating improvement in liver diseases and thus in all types of jaundice.

The heart
As regards disorders to cardiac functions, honey can prove useful in cases such as the following:

Bad cardiac circulation accompanied or not by angina pectoris. For mild cases the use of honey alone is recommended; if the patient is taking digitalis honey can also be taken to reinforce the effect of the drug.

Inflammation of the heart muscle and especially in cases of arrhythmia.

Damage to the heart muscle after infectious illnesses.

For all types of high blood pressure, especially hypertonicity with sympathetic-nervous high blood pressure.

As reinforcement of digitalis therapy.

Preparation for major surgery in patients with circulation problems.

High blood pressure

Rauwolfia can be taken together with honey in order to improve the effects obtained. Honey also acts as a regulating factor in cases of low blood pressure.

Honey can always be taken for cardiac and circulatory problems. There are never any side-effects, except in diabetics.

Allergy

Honey leads to improvement in 90 % of allergy cases. It is usually best to use local honey since this contains the types of pollen causing hay fever. Ingestion leads to immunity.

Abstracts for **The XXXIInd International Congress of Apiculture,** 1989, Bucharest

D.Popeskovic; M. Savovic; M. Blazencic; **Contribution to the Knowledge of the Bioactive Value of Honey.** (Yugoslavia)

A complete survey of the bioactive value of honey - for bees as well as for man - is not available.

In this work one example of comparison, of fatty acid content of four honeys (*Castanea, Robinia, Salvia* and poly flower honey) is given. From this example it results that only *Castanea* honey contains arachidonic acid. The importance of this acid is emphasized because it is a basic chemical skeleton for enzymatic synthesis of prostaglandines in men and other vertebrates. Prostaglandines have extraordinary functions on physiology and pathology of men (cardiovascular diseases, thrombosis, arthritis, gynecology, infections, etc.).

D. Popeskovic; M. Dakic; S. Buncic; P. Ruzic; **A Further Investigation of the Antimicrobial Properties of Honey** (Yugoslavia)

Investigations on the antimicrobial properties of honey is still very much an open issue because modern science has still not yet succeeded in finding answers to many of the questions paused in this field.

Our interest in this subject began some time ago when the selection of an appropriate series of microorganisms was made as test objects for this study, i.e. microbes prevalent in meat hygiene laboratory practice - bacteria with access to the human digestive tract (Popeskovic et al., 1981). Likewise, the existence of both general as well as specific chemical properties respective to individual types of honey must be born in mind.

As an extension of our interest in the bioactive effects of individual types of honey, this paper is a presentation of results obtained upon studying the antimicrobial effects of two types of monofloral honey (i.e. of thistle and pear origin). Eleven species of micro organisms common to laboratory practice in meat hygiene were used as the test objects.

67

Materials and methods

Two types of pure honey were used to test their antimicrobial properties: thistle honey *(Cirsium arvense)* and pear honey *(Pirus domesticus)*. A high level of accuracy was employed in both setting up the experiment and selecting of pure honey so that any possibility of contamination by other antibiotics, or bee contamination with pesticides that could have an antimicrobial effect, were excluded. The following micro organisms were used in the investigation: *B. cereus, B. cereus CCM 869, Pseudomonas, Achromobacter, S. typhimurium, Staph. aureus, Staph. aureus ATCC 6538-P, E. coli, B. mycoides, Sarcina lutea, Str. faecalis.* Additional information regarding methodology can be found in previous papers (Popeskovic et al., 1981).

Results and discussion

Values of the antimicrobial effects of two types of honey (thistle honey and pear honey) are shown in the table.

Comparing the results obtained, it can be noted that thistle honey exhibits twice the inhibitory effect of pear honey on *Staphylococcus aureus.* The effect of thistle honey on the micro-organism *Str. faecalis* is 8 times that of pear honey. On the other hand, as compared to thistle honey, pear honey exhibits a greater inhibitory effect only on the bacterial growth of *Pseudomonas* (2.5 times) and *Achromobacter* (2 times). As regards growth of the remaining species of micro-organisms, both honeys exhibit a similar effect.

Micro-organism	Inhibition zone in mm	
	pear honey	thistle honey
B. Cereus CCM 869	4	2
B. cereus	2	0
Pseudomonas	5	2
Achromobacter	4	2
S. typhimurium	3	3
Staph. aureus	6	12
Staph. aureus ATCC 6538-P	4	3
E. coli	3	2
B. mycoides	5	5
Sarcina lutea	3	3
Str. faecalis	2.5	20

Both chemical composition and the results of the antimicrobial effects of individual types of honey bear testimony to the conclusion that a complete caracterization of the various types of honey has not yet been achieved. The use of various biological tests in obtaining a detailed characterization of individual types of honey has a specific purpose, not only in fostering a better understanding of the biological value of honey for bees and man, but also for lending itself to a more precise evaluation of honey quality.

Ch. Kalman; **Medical Properties of Honey.** (Israel)

Interestingly neither 'modern medicine' nor the public knows how old the use of honey is in the history of medicine.

In the first medical book, in the Ebers papyrus written in approximately 1700 BC, many uses of honey are mentioned: as medicine for internal disease, as external surgical dressing, for burns, ulcers and inflammation of the eyes. The old Egyptians knew the laxative properties of honey (Coso) as a remedy. They also mention wax and black wax (propolis) as medicines. These medical facts about honey were brought to Europe by the Greeks.

According to Greek medicine, honey was a useful remedy for gastric and intestinal disorders, as a pleasant laxative. Respiratory troubles were also noted. The sedative and sopoforic powers of honey are known.

Attieas honey *(Thimus capitatus)* was a highly priced remedy for eye infections and disease. It was used as a surgical dressing. Small pox patients were anointed with honey. It was also used as a vehicle for bitter medicines.

After Aristaeus 'invented' beekeeping, Hippocrates proclaimed that 'Honey causes heat, cleans sores and ulcers, heals carbuncles, and running sores'.

Dioscorides in *Materia Medica* mentions honey, bees wax and propolis. Cornelius Celsius stated in *De Medica* that physicians must heal in a safe, quick and pleasing manner *(Tuto/Cito/et Jucunde)* and all this could be achieved with honey.

The ancient Jews used honey freely (and not marmalade) as medicine. Solomon (in Proverbs 24: 13) said, 'My son, eat thou honey, for it is good'. Jonathon, his son, had his eyes enlightened with the aid of honey.

The Koran recommends honey as an excellent medicine. In ancient China and India, honey was a familiar medicine. The Arab physicians, El Majoussy and El Basry, used honey liberally in their practices.

The name of medicine comes from the Portuguese 'Mellizina'. (In Portuguese Mel = Honey).

Lorand of Carlsbad recommends honey as reconstructive.

Schacht of Wiesbaden claims to have cured many hopeless cases of gastric and intestinal ulcers with honey and without surgery. Professor Koch from Davos, Switzerland, finds that honey has a cholinergic influence on heart muscle disorders· Dr Kosta of Israel discovered the beneficial healing properties on heart muscle insufficiencies.

Chaim Kalman, at the First Symposium of Apitherapy, told of some facts about honey and other bee products, on healing duodenal ulcers, burns and other problems·

Dr Jarvis from Vermont explained the influence of honey on general health and well being.

As far as I know, the gynecological departments of two hospitals in Israel use honey after surgery and another uses it after amputations·

Dr Michael Bulman, M.D, F.R.C.O.S., F.R.C.O.G., obstetric and gynecological surgeon at Norfolk and Norwich Hospital, uses honey as a surgical dressing.

S.E.E. Efem; (1988) **Clinical Observations on the Wound Healing Properties of Honey.** British Journal of Surgery, Vol/Iss/Pg. 3/5 (343-346), ISSN; 0179-0358.

Fifty-nine patients with wounds and ulcers most of which (90 per cent) had failed to heal with conventional treatment were treated with unprocessed honey. Fifty-eight cases showed remarkable improvement following topical application of honey. One case, later diagnosed as Buruli ulcer, failed to respond. Wounds that were sterile at the outset, remained sterile until healed, while infected wounds und ulcers became sterile within one week of topical application of honey. Honey healed wounds rapidly, replacing sloughs with granulation tissue. It also promoted rapid epithelialization, and absorbtion of oedema from around the ulcer margins

From Compact Cambridge MEDLINE 1990.

P, Shambaugh; V. Wothington; J.H. Herbert; **Differential Effects of Honey, Sucrose and Fructose on Blood Sugar Levels.**

It is now recognized that dietary carbohydrate components influence the prevalence and severity of common degenerative diseases such as dental problems, diabetes, heart disease and obesity. Fructose and sucrose have been evaluated and compared to glucose, using glucose tolerance tests, but few such comparisons have been performed for a 'natural' sugar source such as honey. In this study, 33 upper trimester chiropractic students volunteered for oral glucose tolerance testing comparing sucrose, fructose and honey during successive weeks. A 75 -gm carbohydrate load in 250 ml of water was ingested and blood sugar readings were taken at 0, 30, 60, 90, 120 and 240 minutes. Fructose showed minimal changes in blood sugar levels, consistent with other studies. Sucrose gave higher blood sugar readings than honey at every measurement, producing significantly (p less than 0.05) greater glucose tolerance. Honey provided the fewest subjective symptoms of discomfort. Given that honey has a gentler effect on blood sugar levels on a per gram basis, and tastes sweeter than sucrose so that fewer grams would be consumed, it would seem prudent to recommend honey over sucrose.

Mixing Bee Products

Bee products are frequently combined in order to obtain better therapeutic results. One example would be a blend of honey, royal jelly and pollen for the following indications:

1. Safeguarding the health of mothers with babies.
2. Vitamin and mineral deficiency.
3. General stomach, liver and digestive system disorders.
4. Neuroses, insomnia, asthenia and convalescence.
5. Particularly stressful physical activity, hard labour, toxic environment, sports.
6. Disorders of the bronchial tracts.
7. Aging process.
8. In all cases where the rich variety of fortifying substances found in these products can be of importance for human health.

See also 'Propolmel' and 'Propolmel Plus' (page 21)

Apilarnil and Apilarnilprop

Apilarnil contains anabolic hormonal substances which are formed in the male larvae (drone larvae) of the honey bee.

The substance was developed by Nicolae V. Iliesiu, the Romanian specialist who has developed various ways of making bee products suitable for human consumption. Apilarnil is known as the "8th" bee product.

In its structure and biochemical composition Apilarnil somewhat resembles royal jelly, but it has a different potential. The active substance is obtained by homogenizing and filtering the drone larvae. These should be culled at an early stage and they are processed together with the food provided for them. The liquid thus obtained is freeze-dried and sterilized before being made into tablets.

Apilarnil is protected by a large number of patents. It has undergone multiple testing in order to obtain recognition by the appropriate authorities. Patents have been granted not only in Switzerland, Germany, Austria, the USSR and France but also in the USA and England. A European patent has also been granted.

The tests were performed on animals and later experiments with Apilarnil were carried out in hospital in Romania. The results confirm to positive activity of Apilarnil.

Apilarnil contains anabolic hormonal substances (precursors) of the androgen type.

In descending quantitative order Apilarnil contains the following mineral salts: phosphor, potassium, calcium, zinc, manganese, iron, magnesium, copper, sodium. The following vitamins are present: A, beta-carotene, xanthophyll, B1, B2, PP.choline. Amino acids so far identified include: lysine, histidine, arginine, asparaginic acid, teronin, serine, glutaminic acid, proline, glycone, alianine, valine, methionin, isoleucine, tyrosine and phenylalanine.

Of the sixteen amino acids found in Apilarnil, six are essential in that they cannot be synthesized by the organism itself but must be
obtained from an outside source.

Lyophilized Apilarnil is stored in airtight containers at temperatures between 0 and 10 °C. with a shelf life of one year.

Applications of Apilarnil

Apilarnil has many applications in both human and veterinary medicine. In the first place it is used as a pharmaceutical agent in apitherapy. But it is also used as an active ingredient in cosmetics where it activates natural resistance in the body; it is also used in animal husbandry since it causes animals to mate more often and produce greater numbers of offspring.

An important aspect is the use of Apilarnil in gerontology; elderly people's diet and general condition are thereby improved and the brain function is moderately stimulated (in cases of senile dementia).

Apilarnilprop

Apilarnil can be supplemented with propolis, which considerably extends its range of applications. In addition, both preparations complement each other, each augmenting the effect produced by the other. This combined preparation has been given the name Apilarnilprop.

It can be used as a general tonic and biological activator in such conditions as debility, asthenia, malnutrition, post-operative weakness, convalescence, skeletal atrophy, delayed puberty, sexual asthenia, physical or mental exhaustion, premature or accelerated ageing and for conditions requiring a tonic.

The tablets produce no side-effects and can be allowed to melt in the mouth as a treatment of various oral conditions (see chapter on propolis).

Apilarnilprop can be obtained in boxes of 100 tablets packed in blister strips. Each tablet contains 10 mg of lyophilized Apilarnil powder and 16 mg of propolis powder. The usual prescription is 3 - 8 tablets per day. The treatment can be spread over several months, followed by a month's interval before continuing treatment.

Abstract from **The XXIXth International Congress of Apiculture of Apimondia,** Bucharest, 1983.

Voica Foisoreano; N.V. Iliesiu; Maria Floristeanu; Iudit Szabi; (Romania). **Efficiency of the Product Apilarnil in Child's Neuro-Psychomotor Development**

The authors undertook a comparative study on the efficiency of Apilarnil and Apilarnil-prop in the treatment of retardation present in children's neuro-psychomotor development in comparison with the product Encephabol (Enerbol) produced by the Clinics of Neurology from Tg. Mures.

They made observations of 30 children with psychomotor retardation, between 2 and 6 years old, treated with Apilarnil for four months, in comparison with 30 other children with the same characteristics, treated with Encephabol (Enerbol tablets)

Result and observations: 10 days after the start of treatment the children treated with Apilarnil showed an increased memory capacity and attention span, with a net decrease of psychomotor instability. After one month of treatment a relative rapid acquisition of speaking habit had begun, so that after four months the children had already a stock of 20 - 30 words known and showed a real interest in their use.

The children treated with Encephabol showed a much more discrete improvement, which was much slower: the first signs of improvement occurred only after 3 - 4 months of treatment.

List of publications cited

The following is a selection of abstracts from scientific research published over the past ten years. They will give an impression of current research and attitudes towards propolis, bee venom, royal jelly, pollen and honey in different parts of the world as well as the large variety of applications:

König, B., (1986) **Studien zur antivirotischen Aktivität von Propolis (Kittharz der Honigbiene, *Apis mellifera*)**, BSc Dissertation, Hannover.(in German). page 21

D. Grunberger, R. Banerjee, K. Eisinger, E.M. Oltz, L. Efros, M. Caldwell, V. Esteevez and K. Nakanishi (1988); **Preferential Cytoxicity on Tumor cells by Caffeic Acid Phenethyl Ester Isolated from Propolis.** In *Experientia* 44 (1988). This ABSTRACT is printed with kind permission of Birkhäuser Verlag AG, Basel, Switzerland. page 22

Extracts reproduced with kind permission of the Head of the Institute of Microbiology Silesian School of Medicine: Prof. Dr S. Scheller:

Stanislow Scheller, Grazyna Gazda, Wojciech Krol, Zenon Czuba, Alexander Zajusz, Janus Gabrys and Jashovan Shani; (1989) **The Ability of Ethanolic Extract of Propolis (EEP) to Protect Mice against Gamma Irradiation.** In *Zeitschrift für Naturforschung* 44c, 1049-1052 (1989). page 24

S. Scheller, W. Krol, J. Swiacik, S. Owczarek, J. Gabrys, and J. Shani; (1989); **Antitumoral Property of Ethanolic Extract of Propolis in Mice-Bearing Ehrlichh Carcinoma, as Compared to Bleomycin.** page 28

S. Scheller. Willczoks, S. Imielski, W. Krol, J. Gabrys, and J. Shani; **Free Radical Scavenging by Ethanol Extract of Propolis.** page 30

N. Krol, Z. Czuba, S. Scheller, J. Gabrys, S. Grabriec and J. Shani; **Anti-Oxidant Property of Ethanolic Extract of Propolis (EEP) as Evaluated by Inhibiting the Chemiluminescence Oxidation of Luminol.** page 32

Abstracts from The XXXIInd International Congress of Apiculture - 1989, Apimondia, Bucharest.

N.D. Chuhrienko et al. (USSR) - **Complex Treatment of Chronic Bronchitis with Apicultural Products** (p. 281). page 35

Cora Rosenthal et al. (Israel) - **Demonstration of the Inhibitory Effect of Propolis on Microbial Strains** (p. 321). page 35

Yang Ruiyu et al. (China) - **The Effects and the Use of Propolis on Veterinary Medicine** (p. 99). page 35

Abstracts from The XXIXth Congress of Apiculture of Apimondia - 1983, Apimondia, Bucharest.

C. Ciurcaneanu et al. (Romania). **Treatment of Skin and Genital Herpes and of**

Herpes Zoster with Aqueous Propolis Extract and Ointment (p 385) page 36
Ch. Kalman (Israel). **Apitherapy Success in Israel** (p. 401) . page 36

M. Pavlek-Mocan and B. Briski. (Yugoslavia). **Back to Nature Cosmetics using Propolis** (p. 414) . page 36

'Propolis' Apimondia 1978, Bucharest.
N. Baidan; N. Oita; Elena Palos; **Using Propolis in Ophthalmology** (p. 162). page 36

The next output is generated from **Compact Cambridge: MEDLINE 1986 Revised for 1990.**

J. Gabrys; W. Krol; S. Scheller; J. Shani: **Free Amino Acids in Bee Hive Product (Propolis) as Identified by Gas-Liquid Chromatography.** page 37
J. Simuth; J. Trnovsky; J. Jelokova: **Inhibition of Bacterial DNA-dependent RNA Polymerase and Restriction Endonuclease by UV-absorbing Components from Propolis.** page 37
G. Martinez Silveira; A. Gou Gudoy; R. Ona Torriente; Palmer Ortiz MC; Falcon Cuellar **Preliminary Study of the Effects of Propolis in the Treatment of Chronic Gingivitis and Oral Ulceration.** [in Spanish) . page 37
C. Myares; I. Hollands; C. Castaneda; T. Gonzalez.; T. Fragoso; R. Curras; C. Soria; **Clinical Trial with a Preparation Based on Propolis 'Propolisina' in Human Giardiasis.** (Cuba) [in Spanish] . page 38

Abstract from **The XXIXth International Congress of Apiculture - 1983,** Apimondia, Bucharest.

N. Varachiu; Cristina Mateescu; N. Luca; F. Popescu; Gh. Pirvuh: **Experimental and Clinical Studies Concerning the Treatment of Several Periodontal Diseases with Apitherapeutic Products.** page 38

S. Roman: C. Mateescu; E. Palos; **Treatment of Some Ggynecological Diseases with Apitherapeutics.** (International Congress of Apiculture, 1989) page 39

Excerpts from **Journal of the Royal Society of Medicine,** volume 83, March 1990, p. 159-160.

J. M. Grange MSc MD, R. W. Davey MFHom **Antibacterial Properties of Propolis (Bee Glue)** (Department of Microbiology, National Heart & Lung Institute, Dovehouse Street, London SW3 6LY) . page 41

K. Rader, A. Wildfeuer, F. Wintersberger, P. Bossinger and H.-W. Mucke. **Characterization of Bee Venom and its main Components by High-Performance Liquid Chromatography.** In Journal of Chromatography, 408 (1987) 341-348. page 46

Abstract from **The XXIXth International Congress of Apiculture - 1983, Apimondia,** Bucharest.

Elena Palos; Filofteia Popescu; **Use of Bee Venom in antirheumatic Drugs.** (Romania)
page 47

Abstracts from **The XXXIInd International Congress of Apiculture - 1989,** Apimondia, Bucharest.

Shi Bolun, et al; **Clinical Observation on Curative Effect of Freeze dried Royal Jelly
Products on Hyperlipemia and Diabetes.** (China) 50
Xu Ligen et al. **The Efficacies and Mechanisms of Pollen in Anticancers and
Anti-aging.** (China) page 56
Wang Weiyi. **Study on the Digestibility and Absorbtivity of Unbroken-Walled
Pollen.** (China). page 56

Abstracts from **The XXIXth International Congress of Apiculture 1983,** Apimondia, Bucharest.

E. Gheorghieva; V. Vassilev; **Pollen in Treatment of Chronic Liver Lesions.** (Russia)
page 57
E. Gheorghieva; V. Vassileu; **Pollen used in Anaemia.** (Russia) page 57

Abstracts from the **XXXIInd International Congress of Apiculture, - 1989,** Apimondia, Bucharest.

D. Popeskovic; M. Savovic; M. Blazencic; **Contribution to the Knowledge of the
Bioactive Value of Honey.** (Yugoslavia) page 67
D. Popeskovic; M. Dakic; S. Buncic; P. Ruzic; **A Further Investigation of the
Antimicrobial Properties of Honey** (Yugoslavia) page 67
Ch. Kalman; **Medical Properties of Honey.** (Israel) page 68

S.E.E. Efem; (1988) **Clinical Observations on the Wound Healing Properties of
Honey.** British Journal of Surgery, Vol/Iss/Pg. 3/5 (343-346), ISSN; 0179-0358. page 69

Abstract from **Compact Cambridge MEDLINE 1990.**

P, Shambaugh; V. Wothington; J.H. Herbert; **Differential Effects of Honey, Sucrose
and Fructose on Blood Sugar Levels.** page 70

Abstract from **The XXIXth International Congress of Apiculture - 1983,** Apimondia, Bucharest.

Voica Foisoreano; N.V. Iliesiu; Maria Floristeanu; Iudit Szabi; Efficiency of the Product
Apilarnil in Child's Neuro-Psychomotor Development. (Romania) page 74

Bibliography

ARMBRUSTER, Prof. Dr L., Gelée royale (1960), Archiv fur Bienenkunde
1960 p. 1-39 (in german)

Apimondia. La propolis (1975), Apimondia, Bucharest. (Also translated in english: A remarkable hive product: Propolis (1978) Apimondia, Bucharest)

Apitherapie '81: Special issue of 'La revue Francaise d'apiculture'
of the 'Union Nationale de l'Apiculture Francaise (in french):

I **La gelée royale:**
Trente années d'experience, MARY, M.; Les grandes indications de la gelée royale, DONADIEU, Y.; La gelée royale chez les personnes agees, DESTREM, H.

II **Le miel:**
L'Analyse sensorielle des miels, GONNET, M. VACHE, G.; Le miel a-t-il une valeur thérapeutique?, CHAUVIN, R.; Un stage réussi; Les facteurs antibiotiques naturelles présents dans le miel, GONNET, M.; Miel et diabète, DONADIEU, Y.

III **Le pollen:**
Le pollen pratique BONIMOND, J.P.; Pollen et pain d'abeilles: deux cocktails bien faissants, BORNES, G.; Une demande Americaine. Sur le pollen, CHAUVIN, R.; Pollen et prostatisme, DONADIEU, Y.

IV **La propolis:**
Etude du pouvoir bactériostatique de la propolis, COLIN, E.; La propolis en neurologie, MONPLAISIR, M.A.; Propolis et art dentaire, CASTEL, A.; Os et propolis, GIDOIU, Tr., SAFTA, T., PAMBUCCIAN, Gr., PALOS, E., SERBANESCU, M., DOSIUS, L., BLAJA, N.

V **Le venin:**
L'Abeille et son venin, BONIMOMD, J.P.; Recherches sur le venin, URBANECK, R. (arts); Microscopie électronique: de nouveaux univers; Allergie chez les apiculteurs, BOUSQUET, J.; Désensibilisation au venin d'hyménoptères chez les apiculteurs et leur familles, BOUSQUET, J., BONIMOND, J.P and MICHEL, F.B.; Allergie au venin d'hyménoptère, BOUSQUET, J., BONIMOND, J. -P. and MICHEL, F. -B., Allergie aux hyménoptères chez les apiculteurs, une enquète régionale, MICHEL, F.B., AUBRON, J.L., BOUSQUET, J., COULOMB, Y., ALQUIE, M.C. and ROBINET-LEVY, M.; Qu'est-ce-que l'aspi-venin?, EMERIT, A.; Venin et rhumatisme, SERBAN, E.; L'Apithérapie, science de toujours, PERTL, E.; Situation de l'apithérapie aux Etats-Unis, MRAZ, C.; Prévenir l'arthrite, SAINE, J.; Venin d'abeilles et artérite, PARTHENIU, A.

VI **Prise idéale des produits de la ruche:**
Prise idéale des produits de la ruche pour une bonne ou une meilleure santé, DONADIEU, I.

VII **Therapeutiques a base de produits de la ruche associés:**
Propolis et affections rhino-pharyngées, PAUNESOU, C.; Produits de la ruche et thérapie des affections de l'appareil uro-génital, ROMAN, S.-S.; Propolis et affections digestives, IONEL, A.R.; Propolis et affections pulmonaires, IONEL, A.R.; Rétinopathie diabétique et arterite, IONESCO, D. and PARTHENIU, A.;

BANKOVA, V.S., MAREKOV, N.L., Highperformance liquid chromatographic analysis of flavonoids from propolis. Journal of chromatography (1982) 242 (1) 135-143.

BEKENMEIER, H.; BRAUN, W.; FRIEDRICH, E.; KALA, H.; METZNER, J.; SCHNEIDE-WIND, E.; SCHWAIBERGER, R.; WOZNIAK, K.-D. (1973) Mikrobiologisch und klinische Untersuchungen zur Wirksamheit von Propolis. Derm. Mschr. 159(4): 443-449. (in german)

BUMAN, M.D., (1953), M2 Woelm in der Gynaekologie, Praxis 14 (in German)

CAILLAS, A. (1974), La propolis, in:'L'abeille de france et l'apiculteur', March p. 97 (in French)

CAILLAS, A., (1976), Le pollen et les troubles de la prostate, Giens (in french)

CHAUVIN, Prof. Dr R., La valeur diététique des produits de la ruche, miel, gelée royale (1959) Produits pharmaceutiques 4-5-7 (in french)

CHAUVIN, R. (1968) Traite de la biologie de l'abeille [Masson et Cie] Les abeilles et moi [Librairie Hachette edit. 1976] (in french)

CIZMARIK, J. (1976b) Wirkung von Propolis auf Bakterien; Pharmazie 31(9): 656-657 (in German)

CIZMARIK, J., (1976a) Propolis Wirkung auf Hautpilze; Pharmazie 31(1): 55 (in German)

CORSI, M., (1981), A contribution to knowledge about the essential oils of propolis. In Procedings of the XXVIIIth International Congress of Apiculture. Acapulco, 1981. Bucharest, Romania; Apimondia 419-423

CRANE, Eva,. (1975), Honey, a comprehensive study. Heinemann, London

CRANE, Eva., (1980), A book of honey. Oxford University Press

DONADIEU, Y. La gelée royale [Maloine edit., 3d ed. 1978] Le pollen [Maloine edit., 4th ed. 1978] Le miel [Maloine edit., 2d ed. 1978] (in French)
DONADIEU, Y., Propolis in natural therapeutics. Paris, Maloine Editeur S.A. (1983) Ed. 2, 56 pp

ISFELD, Fr., (1968), Pollen als Heilmittel und Kosmetikum. Imkerfreund 2, 1968 (in German)

FOGED, Harley. (1985) Propolis, naturens universalmedicin, Bogans forlag, Denmark (in Danish)

GHISALBERTI, E.L., JEFFERIES, P.R., LANTERI, R., (1977) Potential drugs from propolis.

p. 111-130 of 'Mass spectrometry in drug metabolism'; eds Frigerio, A., Ghisalberti, E.L., New York: Plenum Press

GONNET, M. (1968a) Propriétés phyto-inhibitrices de la colonie d'abeilles *(Apis mellifica L)*. Action de la propolis et de quelques autres produits de la ruche sur la croissance chez Solanum tuberosum. Annls Abeille 11(2): 105-116 (in French)

GONNET, M. (1968b) Propriétés phyto-inhibitrices de quelques substances extraites de la colonie d'abeilles *(Apis mellifica L)*. Action sur la croissance de *Lactua sativa*. Annls Abeille 11(1): 41-47 (in French)

GONNET, M., LAVIE, P. (1960) Action antigerminative des produits de la ruche d'abeilles *(Apis mellifica L.)* sur les graines et les tubercules. C.r. hebd. Seanc. Acad. Sci., Paris 250: 612-614 (in French)

GONNET, T. (1975) Propriétés phyto-inhibitrices de quelques substances extraites de la colonie d'abeilles, in: 'La propolis', Apimondia, p. 70-87, Bucharest (in French)

GORBATENKO, A.G., (1977) Treatment of ulcer patients with a 30 per cent alcohol solution of propolis, Vrach. delo. 3: 22-24 (in russian)

GRITSENKO, V.I., TIKHONOV, A.I., PRYAKHIN, O.R. and PASECHNIK, I.Kh. Photoelectrocolorimetric method of determining a polysacharide preparation from propolis. 'Pharmaceutical Chem. Journal' (1978) p. 579-580

HABERMANN, E., (1972), Bee and wasp venoms, Science, Vol. 177 p. 314 - 324

HEROLD, E., (1970) Heilwerte aus dem Bienenvolk. Munchen: Ehrenwirth Verlag p. 188 (in German)

HEROLD, E., Bienengift bei Herpes im Auge; Imkerfreund, 1964, p. 328 (in German)

HEROLD, E., Bienengift bei alten Leiden; Imkerfreund, 1964, p. 328 (in German)

HEROLD, E., Deckelwachs gegen Asthma; Imkerfreund 1968, p. 200 (in German)

HEROLD, E., Honigwasser als Schlafmittel; Imkerfreund 1964, p. 366 (in German)

HEROLD, E., Kittharz im Kongreszbericht; Imkerfreund 1969, p. 371 (in German)

HEROLD, E., Kittharzsalbe; Imkerfreund 1961 p. 446 (in German)

HILL, Ray., (1977), Propolis. Thorsons publ. ltd, Wellingborough

HLADON, B. et al., (1980) In vitro studies on the cytostatic activity of propolis extracts. Arzneimittel Forschung, 30 (11) 1847-1848. Dept. Pharmacology, Sch. Medicine, Poznan, Poland

JANES, K., BUMBA, V. (1974) Beitrag zur Zusammensetzung des Bienen-harzes (propolis). Pharmazie 29 (8): 544-545 (in German)

JOIRISCH, Prof. Dr N., Weisselfuttersaft und Propolis gegen Grippe; Imkerfreund 1969, p. 11 (in German)

JOIRISCH, Prof. Dr med. N. en RABINOWITSCH, I.M., Propolis - ihre chemische Zusammensetzung und phytocide Wirkung, Imkerfreund 2,1969 (in German)

JUNG, G., Honigmilch - Ernährung bei Frühgeborenen und Neugeborenen, Therapeutische Umschau 7, 1961 (in German)

KARIMOW, S.Ch., Propolis als Medizin; Translation from Ptschelowodstwo by Hartmann; Imkerfreund 1970 nr 8 (in German)

KAAL, J.(1986) Apitherapie, genezing met produkten van bijen *(Apis mellifera)*, Drukkerij Kaal. Amsterdam (in Dutch)

KAAL,J. (1986) Bijen Gezondheidsboekje, Drukkerij Kaal, Amsterdam (in Dutch)

KAAL.J. (1991) Natural Medicine from Honey Bees. Drukkerij Kaal, Amsterdam

KAUL, Dr St., (1967), Erfahrungen mit Bienenhonigloesungen in der Algemeinpraxis, Phys. Med. und Rehabilitation 11, (in German)

KIVALKINA, V.P., (1964) Propolis, its antimicrobial and curative properties. BSc. dissertation (Kazan, 1964) (in russian)

KÖNIG, B., (1986) Studien zur antivirotischen Aktivität von Propolis (Kittharz der Honigbiene, *(Apis mellifera L.)*. BSc. Dissertation Universität Hannover, Germany (in German)

KRAVCHUK, P.A., (1971) Application of propolis for the treatment of chronic subatrophic and atrophic pharyngitis. Kiev (in russian)

LINDENFELSER, L.A. (1967) Antimicrobial activity of propolis, Am. Bee J. 107(3): 90-92; 130-131

MARCHENAIS, PH. (1976) La propolis et ses utilisations 'Abeille de France' nr 591, 5 p (in French)

MARCHENAIS, Ph. (1977) La propolis (1977) Imprimerie Valettoise, Paris (in French)

METZNER, J., BEKEMEIER, H., SCHNEIDEWIND, E., SCHWAIBERGER, R. (1975) Bioautographische Erfassung der antimikrobiell wirksamen Inhaltstoffe von Propolis. Pharmazie 30(12): 799-800 (in German)

MOEBUS, B. (1972) The importance of propolis to honeybees. Br. Bee J. 100: 198-199, 246-248

ORKIN, V.F., (1971) Therapeutic effectiveness of propolis in acute pyrodermatitis. Saratov (in Russian)

PARTIOT (1975) La propolis et la santé, 'l Abeille de France' nr 580 p. 57 (in French)

RIEDACKER, A. (1975) La propolis, in 'Journal du docteur nature' nr 7 libr. Maloine éd, p. 28-31 (in French)

ROBBINS, R.C. (1975) Action in human blood of methyxylated flavones which confer disease resistance on both plants and animals. Concept of a dietary conditioned mechanism of defence against disease. Inst. J. Vit, Nut. Res. 45(1(: 51-60

SCHNEIDEWIND, KALA, LINZER and METZNER (1975) 'On the knowledge of the constituents of propolis' in 'Pharmazie' nr 30 (12) p. 803 (in German)

SCHOLNAST, Chr., Heilung durch Bienengift; Imkerfreund 1967, p. 372 (in German)

SCHWEISHEIMER, Dr W., Honig in der Medizin; Imkerfreund 1968, p.220 (in German)

SHARMA, H, C., SINGH, O.P. Medicinal properties of some lesser known but important bee products. In 'Proceedings of the Second International Conference on Apiculture in Tropical Climates, New Delhi 1980. (1983) 694-702

VILLANUEVA, V.R., BARBIER, M., GONNET, M., LAVIE, P. (1970) Les flavonoides de la propolis. Isolement d'une nouvelle substance bactériostatique: la pinocembrine. Annals Inst. Pasteur. Paris 118(1): 84-87 (in French)

WALRECHT, B.J.J.R. (1962) Over de biologische betekenis van de propolis, 'Biologisch jaarboek' nr 30 p. 253-262 (in Dutch)

WEBER, Dr A., Bienengift als Heilmittel; Imkerfreund 1966, p. 356 (in German)

Index

Ailments

Pathogenic agents

Achromobacter 68
Bacillus alvei 13, 35
Bacillus cereus 42, 68
Bacillus cereus CCM 869 68
Bacillus coli 35
Bacillus larvae 13
Bacillus subtilis 13, 35, 42
Branhamella catarrhalis 42
Candida 39, 40
Candidiasis 40
Corynebacterium spp. 42
Döderlein bacillus 39
Dystrophic diseases 41
Encephalitis viralis 36
Enterobacty clocea 35
Enterococcus spp. 42
Erysipelothrix rhusiophatia 35
Escherichia coli 35-37, 42
Escherichia trychophytes 19
Halitosis 16
Herpes zoster 36
Herpes viruses 22
Listerella monocytogenes 35
Mycobacterium tuberculosis 42
nasopharynx carcinoma cells 28
Osteomyelitis 36
Pasteurella boviseptica 35
Pasteurella suiseptica 35
Proteus bulgaris 13
Pseudomonas 68
Pseudomonas aeruginosa 42
Salmonella abortus-equi 35
Salmonella cholerae Suis 35
Salmonella gallinarum 35
Salmonella paratyphi 35
Salmonella pullorum 35
Salmonella typhi 35
Salmonella typhimurium 35
Sh. Flexner 35
Staphylococci 19
Staphylococcus 39
Staphylococcus aureus 35, 43, 68
Staphylococcus aureus (MRSA) 42
Staphylococcus aureus ATCC 6538-P 68
Staphylococcus epidermis 42
Streptococcus B. 35

Streptococcus equi 35
Streptomyces aureofaciens 37
Toxoplasma gondii 42
Trichomonas 39, 40
Trichomonas vaginalis 42

Authors

Baidan, N. 36
Banerjee, R. 22
Beck, Bodog F. 69
Belfever, B. de 49
Benton, A.W 47
Blazencic, M. 67
Bossinger, P. 46
Briski, B. 36
Bulman, Michael 69
Buncic, S. 67
Caillas 52, 53
Caldwell, M. 22
Castaneda, I. 38
Chauvin 49, 53
Chuhrienko, N.D. 35
Ciurcaneanu, N.C. 36
Crane, E. 67
Cremer, 63
Cuellar, Falcon 37
Curras, R. 38
Czuba, Z. 24, 32
Dakic, M. 67
Davey, R.W. 41
Dioscorides 69
Efros, L. 22
Eisinger, K. 22
Esteevez, V. 22
Floristeanu, Maria 74
Foisoreano, Voica 74
Folen, I.G. 69
Fragoso, T. 38
Gabrys, G. 24
Gabrys, J. 28, 30, 32, 37
Gazda, G. 24
Gheorghieva, E. 57
Gonzalez, T. 38
Gou Gudoy, A. 37
Grabriec, S. 32
Grange, J.M. 41
Grunberger, D. 22
Habermann, E. 47
Hardt, K.L. 47
Herbert, J.H. 70
Herold 52
Hippocrates 69
Hollands, I. 38

Iliesiu, N.V. 74
Imielski, S. 30
Jarvis, D.C. 69
Jelokova, J. 37
Joirisch, N. 53
Kalman, Chaim 36, 69
Kaul, S. 65
Kerkvliet, J.D. 51, 53
Koch, 65
Korsikowski, F.V. 47
Krol, J. 32
Krol, W. 24, 28, 30, 37
Lempt, 65
Lenormand 53
Luca, N. 38
Lulat 42
Markwardt, F. 47
Mateescu, Cristina 38
Morse, R.A. 47
Mucke, H. -W. 46
Myares, C. 38
Nakanishi, K. 23
Neumann, W. 47
Oita, N. 36
Oltz, E.M. 22
Ona Torriente, R. 37
Ortel, S. 47
Owczarek, S. 28
Palmer Ortiz, M.C. 37
Palos, Elena 36, 38, 47
Pavlek-Mocan, M. 36
Pirvu, Gh. 38
Popescu, F. 38, 47
Popeskovic, D. 67, 68
Rader, K. 46
Reiz, K.G. 47
Rietman, 63
Roman, S. 38
Rosenthal, Cora 35
Ruzic, P. 67
Saine, J. 45
Savovic, M. 67
Scheller, S. 24, 28, 30, 32, 37
Shambaugh, P. 70
Shani, J. 24, 28, 30, 32, 37
Shi Bolun 50
Silveira, Martinez 37
Simuth, J. 37

89

Subject index